More information about this series at http://www.springer.com/series/14180

Wireless Networks

Series editor

Tho Le-Ngoc • Ruikai Mai

Hybrid Massive MIMO Precoding in Cloud-RAN

Springer

Tho Le-Ngoc
Department of Electrical
and Computer Engineering
McGill University
Montréal, QC
Canada

Ruikai Mai
Department of Electrical
and Computer Engineering
McGill University
Montréal, QC
Canada

ISSN 2366-1186 ISSN 2366-1445 (electronic)
Wireless Networks
ISBN 978-3-030-02157-3 ISBN 978-3-030-02158-0 (eBook)
https://doi.org/10.1007/978-3-030-02158-0

Library of Congress Control Number: 2018960443

This Springer imprint is published by the registered company Springer Nature Switzerland AG
The registered company address is: Gewerbestrasse 11, 6330 Cham, Switzerland

Preface

Massive multiple-input multiple-output (MIMO), which scales up the number of antennas to the order of tens or even hundreds, promises dramatically improved spectral efficiency and link reliability beyond what can be possibly achieved by conventional MIMO. Recently, in an attempt to overcome technical issues such as hardware complexity and power consumption confronted by the conventional fully digital implementation in the large-scale antenna regime, a hybrid RF-baseband precoding/combining architecture has been proposed. By limiting the number of RF chains, a stage of RF analog precoding/combining is introduced in combination with baseband digital precoding/combining. In this monograph, we explore novel strategies for joint RF-baseband optimization on the assumption of two-timescale channel state information (CSI), which consists of the instantaneous effective CSI and the statistical CSI. Considering the instantaneous effective CSI-based linear baseband precoding/combining for the point-to-point massive MIMO and nonlinear baseband precoding for the multiuser massive MIMO downlink, RF beamforming designs are addressed with respect to the performance metrics of mutual information and mean square error (MSE).

The idea is first examined where channel estimation overhead in point-to-point massive MIMO backhauling is alleviated by reducing the high-dimensional MIMO channel to a low-dimensional RF beam space. This is motivated by the observation that the channel power tends to be concentrated in the eigen-domain as a result of limited scattering. Through a joint selection of the constitutive RF transmit and receive beams, the loss of channel power can be minimized, which thus allows for a near-optimal transmission rate under loose and stringent statistical queueing constraints. When the mechanism of hybrid automatic repeat request (ARQ) is enabled, multiple packet retransmissions are likely to experience independent channel realizations. In addition to the spatial diversity, such time diversity presents another possibility for further performance enhancement in the point-to-point massive MIMO. Building upon the knowledge of the previous failed retransmissions, the hybrid precoder and combiner are sequentially optimized for the current packet retransmission, which results in increased average mutual information compared with those that are oblivious to the diversity in the time dimension.

Without receiver cooperation as in the point-to-point massive MIMO, linear baseband precoding for the massive MIMO downlink suffers severe power loss relative to the capacity-achieving dirty-paper coding in a fully loaded homogeneous multiuser environment, where users require an equal-rate service. This is attributed to the fact that the transmission along the eigen-channel with a vanishingly small magnitude consumes the better part of the transmit power. By taking advantage of the optimization of the extra degrees of freedom in the form of vector perturbation (VP), the proposed use of minimum MSE (MMSE)-VP in hybrid precoding for single-cell massive MIMO remarkably outperforms the perfect CSI-based fully digital linear counterpart. Moreover, factoring the nonlinear perturbation effect in the RF design delivers a superior error performance to the existing solutions that fail to do so. While the single-cell processing can be safely assumed in a multi-cell massive MIMO environment provided that the inter-cell interference (ICI) can be made negligible at the expense of underutilized frequency resources, coordinated nonlinear hybrid precoding between base stations is studied as a spectrally efficient means to realize single-cell processing with universal frequency reuse. In light of the inefficacy to achieve ICI mitigation between cell-edge users in the subspace of spatial correlation, the employment of high-dimensional CSI estimates, as generated by leveraging the effective CSI together with the channel statistics, is examined. Specifically, based on such CSI estimates, RF block diagonalization is shown to effectively limit the adverse impact of ICI and thus create multiple single-cell environments in the RF beam domain. The low-dimensional RF beam domain also reduces the exposure of the baseband to the channel estimation errors and hence lessens the resulting performance degradation.

This monograph is aimed at researchers and engineers interested in the challenges and practical implementation of integrating massive MIMO in the Cloud Radio Access Network of 5G cellular systems.

Montréal, QC, Canada Tho Le-Ngoc
Montréal, QC, Canada Ruikai Mai
September 2018

Contents

List of Acronyms

3GPP	Third-generation partnership project
4G	Fourth generation
ACK	Acknowledgment
ADC	Analog-to-digital converter
AoA	Angle of arrival
AoD	Angle of departure
ARQ	Automatic repeat request
BD	Block diagonalization
BER	Bit error rate
bps	Bit per second
BS	Base station
C-RAN	Cloud radio access network
CoMP	Coordinated multi-point transmission/reception
CSI	Channel state information
DAC	Digital-to-analog converter
dB	Decibel
DFT	Discrete Fourier transform
DPC	Dirty-paper coding
EM	Electromagnetic
FDD	Frequency-division duplexing
IA	Interference alignment
ICI	Inter-cell interference
i.i.d.	Independently and identically distributed
KKT	Karush-Kuhn-Tucker
LNA	Low-noise amplifier
LS	Least squares
LTE	Long-term evolution
MAC	Multiple-access channel
MIMO	Multiple-input multiple-output
MISO	Multiple-input single-output
ML	Maximum likelihood

MMSE	Minimum mean square error
mmWave	Millimeter wave
MRC	Maximum ratio combining
MRT	Maximum ratio transmission
MSE	Mean square error
MU-MIMO	Multi-user MIMO
NACK	Negative ACK
OFDM	Orthogonal frequency-division multiplexing
PA	Power amplifier
PAS	Power azimuth spectrum
PD	Positive definite
PSD	Positive semi-definite
QAM	Quadrature amplitude modulation
QoS	Quality of service
RF	Radio frequency
RZF	Regularized zero forcing
SDMA	Space-division multiple access
SDR	Semi-definite relaxation
SINR	Signal-to-interference-plus-noise ratio
SLNR	Signal-to-leakage-plus-noise ratio
SNR	Signal-to-noise ratio
SU-MIMO	Single-user MIMO
SVD	Singular value decomposition
TDD	Time-division duplexing
THP	Tomlinson-Harashima precoding
ULA	Uniform linear array
UPA	Uniform planar array
UT	User terminal
VP	Vector perturbation
ZF	Zero forcing

List of Symbols

$\mathbf{a}_{\mathrm{r}}\left(\theta^{\mathrm{r}}, \phi^{\mathrm{r}}\right)$	Receive array response to azimuth AoA θ^{r} and elevation AoA ϕ^{r}
$\mathbf{a}_{\mathrm{t}}\left(\theta^{\mathrm{t}}, \phi^{\mathrm{t}}\right)$	Transmit array response to azimuth AoD θ^{t} and elevation AoD ϕ^{t}
B	Channel bandwidth in Hz
$\mathbf{d}$	Data streams
$\mathcal{D}_{\mathrm{t}}, \mathcal{D}_{\mathrm{r}}$	DFT codebook for transmission/reception
d_{λ}	Inter-antenna spacing normalized by the wavelength λ
$\mathbf{E}$	MSE matrix
$\mathbf{F}_{\mathrm{BB}}, \mathbf{F}_{\mathrm{BB},m}$	Baseband digital precoder
f_{c}	Carrier frequency
$\mathbf{F}_{\mathrm{RF}}, \mathbf{F}_{\mathrm{RF},m}$	RF analog precoder
$f_{\tau}\left(\cdot\right)$	Modulo-τ operation
$\mathbf{H}, \mathbb{H}$	MIMO channel matrix
M	Number of packet retransmissions in hybrid ARQ systems
$\mathbf{n}$	Zero-mean, circularly symmetric, spatially white Gaussian noise
N_{cl}	Number of scattering clusters
n_{r}	Number of receive RF chains
N_{r}	Number of receive antennas
N_{ray}	Number of paths per cluster
N_{s}	Number of spatially-multiplexed data streams
N_{t}	Number of transmit antennas
n_{t}	Number of transmit RF chains
P_{t}	Transmit power constraint
$\mathbf{q}, \mathbf{a}$	Perturbation vector
$\mathbf{s}$	Signal before linear baseband precoding
$\hat{\mathbf{s}}$	Signal estimate after linear equalization
T	Channel coherence time in terms of channel uses
$\mathbf{W}_{\mathrm{BB}}, \mathbf{W}_{\mathrm{BB},m}$	Baseband digital combiner
$\mathbf{W}_{\mathrm{RF}}, \mathbf{W}_{\mathrm{RF},m}$	RF analog combiner
$\mathbf{x}$	Transmit signal
$\mathbf{y}$	Receive signal
σ_n^2	Noise variance
σ_e^2	Variance of channel estimation errors

List of Notations

- Column vectors and matrices are denoted by lower and upper boldface letters, respectively
- $\mathbb{R}$ and $\mathbb{C}$ denote the set of real and complex numbers, respectively
- $\mathrm{tr}\,(\mathbf{A})$, $\mathbf{A}^*$, $\mathbf{A}^{\mathrm{H}}$, and $|\mathbf{A}|$ denote the trace, conjugation, conjugate transposition, and determinant of $\mathbf{A}$, respectively
- $\mathbf{0}$ and $\mathbf{1}$ are the all-zero vector and all-one vector, respectively
- $\mathbf{I}_N$ is the $N \times N$ identity matrix
- $\otimes$ is the Kronecker product, and $\odot$ represents the Hadamard product
- $\mathrm{vec}(\mathbf{A})$ is the vectorizing operator on an entire matrix, i.e., for $\mathbf{A} \in \mathbb{C}^{N\times N}$, $\mathrm{vec}(\mathbf{A}) = [a_{1,1}, \ldots, a_{N,1}, \ldots, a_{1,N}, \ldots, a_{N,N}]^{\mathrm{T}}$, while $\mathrm{vec}_{\mathrm{l}}(\mathbf{A})$ performs vectorization on the elements below the main diagonal of $\mathbf{A}$, i.e., $\mathrm{vec}_{\mathrm{l}}(\mathbf{A}) = [a_{2,1}, \ldots, a_{2,N}, \ldots, a_{N,N-1}]^{\mathrm{T}}$
- $\angle(\cdot)$ retrieves the angle of a complex number in radians
- $\mathrm{diag}\,\{a_1, \ldots, a_N\}$ denotes an $N \times N$ diagonal matrix with $a_1, \ldots, a_N$ on the diagonal, while $\mathrm{diag}\,\{\mathbf{A}_1, \ldots, \mathbf{A}_N\}$ is a block diagonal matrix with $\mathbf{A}_1, \ldots, \mathbf{A}_N$ as the block diagonal elements
- $\mathrm{diag}(\mathbf{A}) = [a_{1,1}, \ldots, a_{N,N}]^{\mathrm{T}}$ retrieves the N diagonal elements of $\mathbf{A} \in \mathbb{C}^{N\times N}$
- $\lambda_{\mathbf{A},[i]}$ denotes the ith largest eigenvalue of $\mathbf{A}$
- $\mathbf{A} \succeq 0$ indicates that $\mathbf{A}$ is positive semi-definite
- $[\mathbf{A}]_{i,j}$ denotes the (i, j)th element of $\mathbf{A}$
- $\mathbf{A}^{i_1,\ldots,i_P}_{j_1,\ldots,j_Q}$ is a $P \times Q$ submatrix formed by the rows $\{i_1, \ldots, i_P\}$ and columns $\{j_1, \ldots, j_Q\}$ of $\mathbf{A}$ in increasing order
- $(x)^+ = \max(x, 0)$
- $\jmath = \sqrt{-1}$ is the imaginary unit
- $\Re\{\cdot\}$ and $\Im\{\cdot\}$ denote the real and imaginary part of a complex number/vector, respectively
- δ_{ij} is the Kronecker delta function which is unity for $i = j$ and zero otherwise
- $\mathbb{E}_{\mathbf{x}}\,[\cdot]$ takes expectation over $\mathbf{x}$

- $\mathcal{CN}\left(0, \sigma_n^2\right)$ denotes zero-mean circularly symmetric white Gaussian distribution with variance σ_n^2
- $\mathbb{S}_+^N$ denotes the set of $N \times N$ positive semi-definite matrices and $\mathbb{S}_{++}^N$ represents the set of $N \times N$ positive definite matrices
- $\|\cdot\|_\mathrm{F}$ defines the Frobenius norm, and $\|\cdot\|_2$ is the Euclidean norm

Chapter 1
Introduction

1.1 Scaling up MIMO Communications

According to the Cisco visual networking index [1], mobile data traffic has grown 18-fold over the past 5 years to reach 7.2 exabytes per month at the end of 2016, of which 4G Long-Term Evolution (LTE) traffic accounted for 69%. This trend has largely been driven by the increased use of smart phones and tablets as well as by the emergence of services such as internet of things and machine-to-machine communications. As the mobile connections continue the rapid growth, which is accompanied by the demand of high-definition video streaming, augmented reality, and virtual reality, a rethink of the current architecture of wireless communications systems has become more important than ever.

In comparison with its wireline counterparts, the wireless medium has two distinctive features: (1) the broadcast nature, and (2) the presence of channel fading. One immediate consequence of the broadcast nature is that different links tend to interfere with each other when operating simultaneously over the same time-frequency signaling resources. On the other hand, as signals in the form of electromagnetic waves travel from the source to the destination, they interact with objects in the propagation environment, and experience reflection, diffraction, and scattering. As a result, multiple copies of the same signal arrive superimposed at the receiver. Collectively, the superposition leads to random fluctuation in the received signal strength. Because of such small-scale channel fading phenomena, the efficiency and reliability of wireless systems are highly sensitive to the changes in the propagation environment, which tends to limit the scope of its application. In the past 15 years, the introduction of multiple-input multiple-output (MIMO) technology has fundamentally changed the landscape of wireless communications. Essentially, by equipping the transmitter/receiver with adaptive antenna arrays, directional transmission/reception, also known as transmit/receive beamforming, is enabled by sophisticated signal processing techniques [2, 3]. The physical underpinning is that by shaping the signal emitted from each antenna adaptively

T. Le-Ngoc, R. Mai, *Hybrid Massive MIMO Precoding in Cloud-RAN*,
Wireless Networks, https://doi.org/10.1007/978-3-030-02158-0_1

to the channel conditions, the wavefronts will be collectively superimposed in a constructive fashion at the intended receiver but destructively at other receiver locations. This makes it possible to mitigate the adverse impact of channel fading and reduce the interference to other co-existing pairs of links. The end result is that we have witnessed the drastically improved quality of service (QoS) in terms of data rate and end-to-end latency in modern wireless communication systems. For example, the ongoing standardization efforts by LTE introduced downlink MIMO techniques as a physical-layer enhancement starting from Release 10, which among others serve as a key enabler to fulfill the objective of a maximum spectral efficiency of 30 bps/Hz, a major upgrade from 16 bps/Hz in Release 8 with single-antenna transmission/reception [4].

Conceptually, by adjusting the antenna weights, beamforming techniques attempt to focus the transmit/receive power towards the desired spatial direction, which is characterized by the 3-dB beamwidth of the main lobe. The narrower the main lobe is, the higher the beamforming gain (or array gain) is. However, given the size of an antenna array, physics dictates a fundamental trade-off between the array gain and the leakage power in the form of side lobes and back lobe. The implication is that while it is desirable to realize focused transmission/reception by narrowing the main lobe to a maximum extent, unwanted power radiated from the side lobes and back lobe will inevitably become increasingly significant. This unfortunately offsets the array gain and the effectiveness of interference suppression of beamforming. Naturally, such a trade-off motivates the consideration of scaling up the conventional MIMO by an order of tens or even hundreds, which is known as massive MIMO [5]. In doing this, we are able to reap the benefits of conventional MIMO on a large scale.

As dictated by Shannon's channel capacity formula, the highest data rate is determined by two factors: (1) the received signal-to-noise ratio (SNR), and (2) the operating bandwidth [6]. While the former can be boosted by signal processing techniques such as beamforming, the latter relies on the availability of the radio frequency spectrum. The next-generation wireless systems have envisioned data rates on the order of gigabits per second, which, subject to a practical power constraint, is impossible to deliver without wideband channels [7, 8]. Traditionally, thanks to the favorable propagation behavior, most consumer wireless systems have been designed to operate in the microwave (sub-6 GHz) band. As the frequency spectrum under 6 GHz becomes densely occupied over time, the millimeter-wave (mmWave) band (30 to 300 GHz) has attracted the attention of the research community in search of widely available spectrum resources [9]. This is however not without its challenges. When modulated to an mmWave carrier frequency, the electromagnetic waves become highly sensitive to path loss and blockage. In other words, without further physical-layer enhancements, mmWave communications would be restricted to short-range, line-of-sight applications. Fortunately, as the form factor of an antenna array is proportional to the wavelength, operating over the mmWave frequency band provides the possibility of integrating a large-scale antenna array into a much smaller area than its microwave counterparts. For example, at a carrier frequency of 60 GHz, the form factor of an 8×8 planar antenna array is only 4 cm^2. Accordingly, the issues of severe path loss and blockage can be effectively addressed by massive MIMO beamforming, which paves the way for the deployment of mmWave in cellular networks [10].

Over an assigned set of time-frequency resources, expansion of the network capacity in terms of area spectral efficiency, can be realized by admitting as many users as possible. For this purpose, one straightforward approach is to aggressively densify the deployment of base stations (BSs) with universal frequency reuse [11]. Such infrastructure densification reduces the probability of a large number of users competing for resources within a cell. The resulting decrease of the cell radius also enables low-power transmission, and hence increased energy efficiency. Furthermore, load balancing allows each cell to handle a higher data traffic demand from the associated users in service. However, before fully realizing the benefits of BS ultra-densification, severe inter-cell interference, among other performance limiting factors, needs to be appropriately addressed. Besides, in areas lacking in infrastructure, load balancing cannot be easily implemented via fiber-optic cable connections. For these challenges, one promising solution is to equip the BSs with large-scale antenna arrays. The rich degrees of freedom (DoFs) present a great flexibility of beamforming to achieve spatial separation at the transmitter and receiver. Through coordination and cooperation across interfering cells, inter-cell interference suppression and management become possible. Besides, the BSs can communicate with each other directly through the air interface without requiring an extensive investment of backhaul infrastructure. The high array gain of massive MIMO together with a large amount of idle mmWave frequency spectrum makes it possible to handle a wide range of QoS prescriptions.

1.2 Massive MIMO: Architecture, Challenges, and Opportunities

1.2.1 Hardware Architecture

From the perspectives of signal processing and hardware implementation, beamforming can be categorized as (1) RF analog beamforming, (2) baseband digital beamforming, and (3) hybrid RF-baseband beamforming.

The mechanism of RF transmit/receive analog beamforming is illustrated in Fig. 1.1. For transmit beamforming, the complex baseband signal is first modulated to the desired RF carrier frequency. The amplitude and phase of the continuous RF signal are then respectively adjusted by a power amplifier (PA) and phase-shifter before being radiated from each antenna element. The RF analog beamformer enjoys low hardware complexity with a single RF chain employed. Such a simple architecture lends itself to large-scale antenna arrays. On the downside, a single

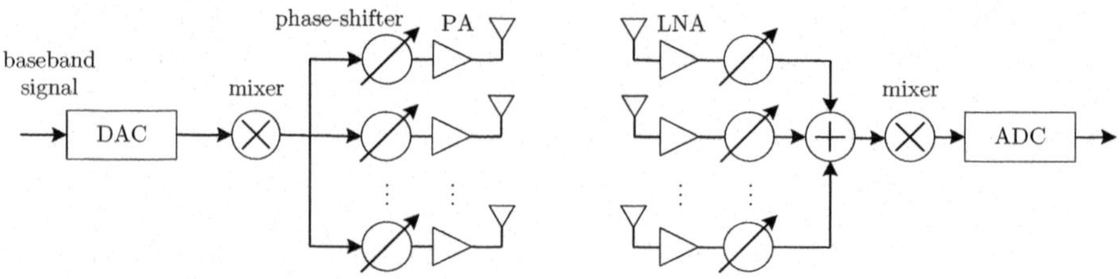

Fig. 1.1 RF analog transmit/receive beamforming

RF chain disables spatial multiplexing, and thus restricts the usage of RF analog beamforming to low-rate transmission/reception. Furthermore, RF phase-shifters generally have a finite resolution, and therefore limited flexibility of generating beamforming directions. Finally, for adaptive RF beamforming, it is cumbersome to update the antenna weights in sync with fast channel variations.

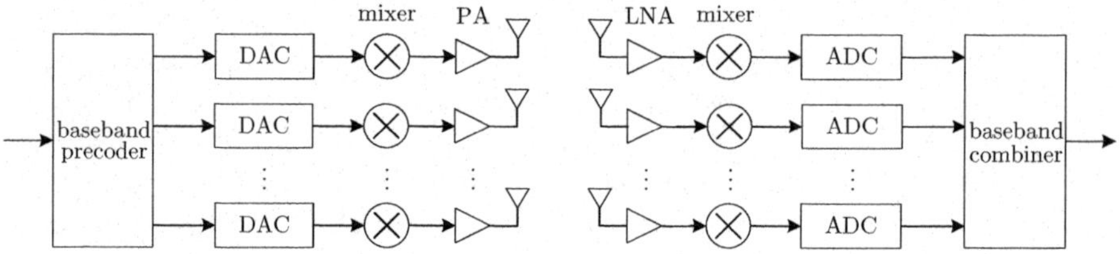

Fig. 1.2 Baseband digital precoding/combining

Given the limitations of RF analog beamforming, one popular alternative is to adjust the antenna weights at baseband, i.e., baseband digital transmit/receive beamforming, as demonstrated in Fig. 1.2. Differently from RF analog beamforming which manipulates the amplitude and phase of a continuous RF signal, digital beamforming modifies the in-phase (I) and quadrature (Q) components of a complex baseband signal. Enabled by powerful digital signal processing techniques, baseband digital beamforming is more flexible and responsive to channel variations. In addition, spatial multiplexing now becomes a viable option. Such advantages over RF analog beamforming come at the expense of a dedicated RF chain for each antenna element. For accurate uploading/downloading of baseband signals to/from the antenna domain, high-performance analog-to-digital converters (ADCs)/digital-to-analog converters (DACs) in terms of precision and speed are needed. Unfortunately, such high-performance ADCs/DACs tend to be power-hungry and costly, which consequently limits the number of RF chains in deployment. The limitations of the fully digital beamforming architecture in terms of hardware complexity, cost, and power consumption make it quickly become impractical when scaling up the antenna array.

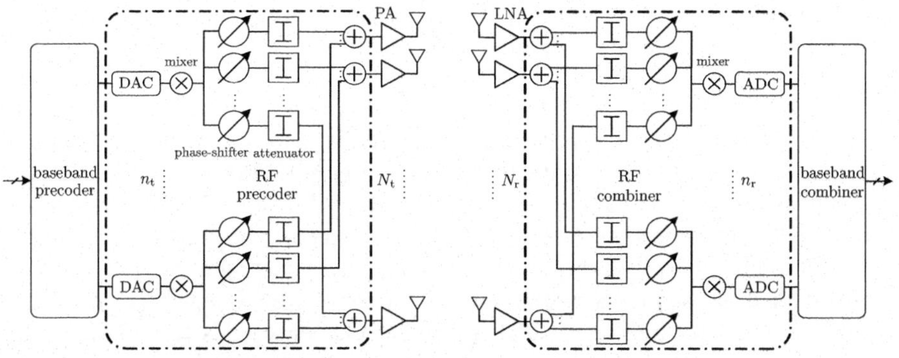

Fig. 1.3 Hybrid RF-baseband precoding/combining

Weighing the pros and cons of the RF analog and baseband digital structures, a potential solution to strike a trade-off between performance and practicality in the implementation of massive MIMO is hybrid RF-baseband precoding/combining, as illustrated in Fig. 1.3. The hybrid RF-baseband beamforming architecture is a natural consequence from driving a large-scale antenna array with a reasonably small number of RF chains, and is derived from combining RF analog beamforming with baseband digital beamforming.

1.2.2 Challenges and Opportunities

Massive MIMO was first envisioned for time-division duplexing (TDD) systems [12] in favor of its relatively low overhead of channel estimation. Provided that the downlink and uplink transmissions take place over the same channel coherence block, the downlink channel can be theoretically treated as the transpose of the uplink counterpart. Accordingly, via uplink channel training, the downlink channel state information (CSI) can be inferred and leveraged for spatial multiplexing and interference management. In doing this, the overhead of channel estimation scales with the number of users instead of the number of antennas at the BS as in typical frequency-division duplexing (FDD) systems. However, in practice, due to the asymmetry between RF chains at the BS and user terminals, hardware calibration is required before such channel reciprocity can be safely assumed [13]. Moreover, the issue of pilot contamination, if not addressed properly, imposes a fundamental limit on the system performance [5]. Specifically, because of the limited availability of orthogonal pilot sequences, it is inevitable that the same pilot sequence be used across multiple cells. Against the backdrop of universal frequency reuse, the reuse of pilot sequences is likely to introduce out-of-cell interference into the uplink channel estimation, and call into question the accuracy of the acquired channel estimates.

When the number of antennas is asymptotically large, the independently and identically distributed (i.i.d.) Rayleigh fading channels experience the so-called favorable propagation condition [5]. For example, for the multi-user multiple-input single-output (MISO) downlink, the channel vector with respect to each user becomes increasingly orthogonal to each other as the number of transmit antenna grows. In other words, the inter-user interference which is present in the limited antenna regime vanishes in the large antenna regime, and the multi-user channel degenerates into multiple non-interfering single-user MISO channels [14, 15]. In this case, it suffices to employ simple single-user MIMO processing such as maximum ratio transmission. A similar observation can also be made for the multi-user single-input multiple-output (SIMO) uplink, where maximum ratio combining can be applied for single-user detection. For massive MIMO operating over the microwave frequency band, the number of array elements tends to be limited by a not-so-small form factor, and it is thus difficult to justify the validity of such an ideal propagation condition. On the other hand, while the small form factor at the mmWave frequency band makes it possible to integrate a large number of antenna elements, rich scattering is unlikely to be observed [16], and renders invalid the assumption of i.i.d. Rayleigh fading. Therefore, sophisticated multi-user MIMO processing strategies still need to be devised with attention to the potentially high computational complexity due to the high-dimensional MIMO channels.

Although the TDD mode has been commonly accepted as a more, if not the only, viable option for massive MIMO, it is still of practical interest to evaluate the potential of integrating massive MIMO with the FDD mode which is dominant among existing systems. For closed-loop FDD, CSI acquisition at the transmitter relies on the mechanism of channel training and feedback, and the associated signaling overhead is generally on the order of the array dimension. When directly applied to the high-dimensional channel matrix of massive MIMO, the feasibility of such a conventional approach becomes questionable because of its demand of excessive time-frequency resources. Besides, the large volume of CSI feedback, whether it is digital or analog, might render the CSI estimates obsolete. Fortunately, there exists uplink-downlink channel reciprocity in terms of the channel spatial correlation even when the uplink and downlink frequency bands are separated apart [17]. In this case, the downlink channel correlation can be estimated by uplink channel training. Such partial CSI in turn serves as the basis for devising channel estimation schemes with reduced overhead.

While a high-dimensional MIMO channel can be created by equipping the transmitter/receiver with large-scale antenna arrays, it is in fact the eigen-directions and eigen-gains of the channel that are most relevant to the effectiveness of spatial multiplexing, and thus the throughput. In the absence of a rich scattering environment, transmit/receive correlation tends to be present and thus decrease the number of data streams that can be physically supported. For example, for a typical macro-cell scenario, it is unlikely for a highly-mounted BS to be surrounded by abundant scatterers, and therefore it is expected that the BS antennas experience spatial correlation as a result of limited scattering. The phenomenon of limited scattering is actually inherent in mmWave propagation [16]. On the upside, such rank deficiency of the high-dimensional MIMO channel physically justifies the sufficiency of using a limited number of RF chains. Besides, the channel spatial correlation, which tends to vary on a much slower timescale than the instantaneous channel realization, can be exploited to facilitate the design of low-complexity channel estimation schemes in FDD massive MIMO systems.

1.3 Theme and Organization of Monograph

With a promise to deliver key performance indicators beyond what is possible with conventional MIMO, massive MIMO is widely deemed as a essential component for the next-generation wireless communications systems. In fact, the technique of massive MIMO has been embraced by the industrial standardization efforts, e.g., starting from 3GPP LTE Release 13 [4]. Whether or not massive MIMO can be realized to its full potential depends on how successfully the aforementioned technical challenges will be addressed. The theme of this monograph is to explore practical hybrid RF-baseband precoding/combining solutions in consideration of system limitations such as hardware complexity, channel estimation overhead, and channel spatial correlation. Novel techniques of joint RF-baseband optimization are devised with respect to the performance metrics of mutual information and MSE for various application scenarios. The hybrid RF-baseband architecture, which is derived from limiting the number of RF chains attached to a large-scale antenna array, bears both resemblance to the fully digital architecture and unique challenges in itself. On one hand, by treating as the effective channel the cascade of the RF analog stage with the high-dimensional MIMO channel, a wealth of conventional MIMO solutions can be readily exploited for the baseband digital stage. On the other hand, the introduction of the RF analog stage presents a potential avenue towards decreased computational complexity and channel estimation overhead, which necessitates novel methods of performance optimization subject to various practical constraints.

In that the hybrid RF-baseband structure is a suboptimal alternative to the conventional fully digital implementation, we first study joint RF-baseband optimization at the transmitter and receiver to minimize the performance loss for point-to-point massive MIMO systems. For the case of wireless backhauling where the objective of mutual information subject to statistical queueing constraint is of interest, we derive the optimal structures of RF precoder and combiner, and shed light on how to jointly select the transmit-receive beamforming directions such that a near-optimal performance can be achieved. When packet retransmissions are enabled through hybrid automatic repeat request (ARQ), we analytically characterize the optimal baseband precoder which improves the mutual information by exploiting the temporal diversity and incorporating the effect of RF precoding on power loading. We also devise a hybrid combining scheme that decouples itself from the hybrid precoding optimization, and reduces the overall storage requirement and computational complexity for each ARQ round of retransmission. Motivated by the severe power loss suffered by the linear precoding schemes in fully-loaded multi-user systems, we explore the combination of nonlinear minimum mean square error (MMSE)-vector perturbation (VP) with hybrid precoding based on the two-timescale CSI in a single-cell environment. We derive a non-iterative approach to incorporate the effect of nonlinear VP into RF precoding optimization, and demonstrate the performance advantage of such a joint design. Equipped with this insight, we further extend the nonlinear hybrid precoding solution to a multi-cell scenario with emphasis on the cell-edge users, which typically represent the limiting performance of a cellular network. We investigate how to exploit the two-timescale CSI more effectively such that superior RF precoding solutions to the channel spatial correlation-based state of the arts can be generated. This work demonstrates the flexibility and advantage of adapting RF precoding to the imperfect CSI produced by channel tracking. It is also concluded that the presence of the RF stage effectively reduces the exposure to channel estimation errors at baseband, which hence alleviates the negative impact of such errors on VP.

The remainder of this monograph is organized as follows.

Chapter 2 presents the preliminaries and recent development of precoding/combining design in the context of both conventional and massive MIMO, which lay the foundation for the hybrid precoding/combining techniques studied in the subsequent chapters.

Chapter 3 is concerned with a joint design of two-timescale hybrid precoding and combining to maximize the effective rate for massive MIMO-enabled wireless backhauling. Specifically, the RF analog precoding/combining is adaptive to the statistical CSI while the digital baseband precoding/combining is updated with the instantaneous effective CSI. In consideration of the conventional MIMO solutions at baseband, the issues of joint RF precoding-combining with both nonuniform- and constant-modulus elements are addressed. In the former case, we derive the optimal RF solution structures, which lead to a problem formulation of combinatorial eigenmode selection. Such an NP-hard problem is solved to near-optimality by semi-definite relaxation. In view of the additional difficulty posed by the nonconvex modulus constraint, we exploit the problem structure to construct the constant-modulus design from the nonuniform-modulus solution, which is cast as a problem of joint matrix approximation and solved by low-complexity Jacobi-like algorithms. Numerical results show that under loose and stringent delay-outage constraints, the two-timescale hybrid designs deliver effective rates comparable with other perfect CSI-based state-of-the-art baselines.

Chapter 4 studies a design of perfect CSI-based hybrid precoding and combining for massive MIMO ARQ systems. The proposed progressive hybrid precoder and combiner aim at performance enhancement through exploiting the time diversity inherent in hybrid ARQ where the same packet is retransmitted in the event of decoding failure. In view of the intractability of joint optimization of the hybrid precoder and combiner, we heuristically assume that the linear Wiener filter is perfectly realizable by the hybrid RF-baseband combiner at the receiver, and thus decouple the precoding from the combining design. The knowledge of the previous retransmissions is incorporated into the joint RF-baseband optimization, which is sequentially performed to increase the average mutual information. In particular, for each round of retransmission, we choose the RF precoder either from the transmit array response vectors or from a discrete Fourier transform (DFT)-based codebook. Based on the chosen RF precoder, the optimal baseband precoder is analytically shown to be a function of the generalized eigenvectors of the effective channel and the RF precoder for the current ARQ round with power allocation determined by the precoding solutions from the previous and current transmissions. To ensure the feasibility of the decoupled precoder-combiner design, the proposed two-step design technique is further applied to derive a hybrid RF-baseband combiner as an approximation of the optimal linear digital solution. Illustrative examples verify the efficacy of the proposed progressive hybrid solutions by performance comparison with various baselines. In addition, the performance of the proposed design is numerically examined in the presence of quantized RF elements.

Chapter 5 explores a joint design of two-timescale hybrid precoding with MMSE-VP for multi-user massive MIMO systems. Users are assumed to be geographically clustered, where each cluster of users experiences the same transmit spatial correlation. Considering the perfect effective CSI-based MMSE-VP at baseband, the statistical CSI-based RF precoder designs are formulated as orthonormality-constrained stochastic optimization problems. For single-cluster transmission, RF eigen-beamforming is shown to be optimal. In multi-cluster scenarios, due to the lack of closed-form characterization of the objective functions, mathematically tractable lower bounds are proposed and numerically optimized by trust-region Newton methods on Riemannian manifolds. In addition, the design of DFT codebook-based RF beam selection is addressed. By recognizing the objective functions as a difference of increasing functions, branch-reduce-and-bound techniques are developed to find the globally optimal solutions at reduced computational complexity. Simulation results illustrate that the proposed nonlinear hybrid schemes deliver a superior bit error rate to other state-of-the-art baselines. The effectiveness of the suboptimal DFT codebook-based RF solutions is also verified.

Chapter 6 extends the idea of nonlinear hybrid precoding with MMSE-VP to multi-cell massive MIMO systems. Two-timescale CSI is assumed, which consists of noisy observations of the short-term RF-beamformed channel, and the perfect knowledge of the long-term channel temporal and spatial correlation. By exploiting the low-dimensional effective CSI, we propose to estimate the instantaneous realization of the high-dimensional MIMO channel via Kalman filtering. The CSI estimate is then utilized for RF beamforming in conjunction with centralized and distributed baseband MMSE-VP, respectively. Specifically, robust baseband solutions are first derived by alternating optimization, where a Jacobi update of the dual formulation is proposed for reduced computational complexity. By abstracting the effect of nonlinear baseband precoding, RF beamforming is separately formulated as a solution to balance the error performance with the accuracy of channel tracking. To solve such nonconvex problems, we develop gradient descent search algorithms based on Cayley transformation. Numerical examples confirm the usefulness of adapting hybrid precoding to the imperfect CSI from channel tracking in terms of its appreciable error performance gain over other channel correlation-based methods. Moreover, the resilience of the proposed solution to the channel estimation errors is verified.

Chapter 7 gives concluding remarks and provides suggestions for future investigations.

References

1. Cisco, "Cisco visual networking index: Global mobile data traffic forecast update, 2016–2021," Tech. Rep., Feb. 2017.
2. L. C. Godara, "Applications of antenna arrays to mobile communications, Part I: Performance improvement, feasibility, and system considerations," *IEEE Proceedings*, vol. 85, no. 7, pp. 1031–1060, Jul. 1997.
3. ——, "Applications of antenna arrays to mobile communications, Part II: Beam-forming and direction-of-arrival considerations," *IEEE Proceedings*, vol. 85, no. 8, pp. 1195–1245, Aug. 1997.
4. E. Dahlman, S. Parkvall, and J. Sköld, *5G NR: The Next Generation Wireless Access Technology*. Cambridge, MA, USA: Academic Press, 2018.
5. F. Rusek, D. Persson, B. K. Lau, E. G. Larsson, T. L. Marzetta, O. Edfors, and F. Tufvesson, "Scaling up MIMO: Opportunities and challenges with very large arrays," *IEEE Signal Process. Mag.*, vol. 30, no. 1, pp. 40–46, Jan. 2013.
6. D. Tse and P. Viswanath, *Fundamentals of Wireless Communications*. New York, NY, USA: Cambridge University Press, 2005.
7. J. G. Andrews, S. Buzzi, W. Choi, S. V. Hanly, A. Lozano, A. C. K. Soong, and J. C. Zhang, "What will 5G be?" *IEEE J. Sel. Areas Commun.*, vol. 32, no. 6, pp. 1065–1082, Jun. 2014.
8. R. W. Heath, N. González-Prelcic, S. Rangan, W. Roh, and A. M. Sayeed, "An overview of signal processing techniques for millimeter wave MIMO systems," *IEEE J. Sel. Topics Signal Process.*, vol. 10, no. 3, pp. 436–453, Apr. 2016.
9. Z. Pi and F. Khan, "An introduction to millimeter-wave mobile broadband systems," *IEEE Commun. Mag.*, vol. 49, no. 6, pp. 101–107, Jun. 2011.
10. W. Roh, J. Park, B. Lee, J. Lee, Y. Kim, J. Cho, K. Cheun, and F. Aryanfar, "Millimeter-wave beamforming as an enabling technology for 5G cellular communications: Theoretical feasibility and prototype results," *IEEE Commun. Mag.*, vol. 52, no. 2, pp. 106–113, Feb. 2014.
11. M. Kamel, W. Hamouda, and A. Youssef, "Ultra-dense networks: A survey," *IEEE Commun. Surveys Tuts.*, vol. 18, no. 4, pp. 2522–2545, Fourth Quarter 2016.
12. T. L. Marzetta, "Noncooperative cellular wireless with unlimited numbers of base station antennas," *IEEE Trans. Wireless Commun.*, vol. 9, no. 11, pp. 3590–3600, Nov. 2010.
13. G. Smith, "A direct derivation of a single-antenna reciprocity relation for the time domain," *IEEE Trans. Antennas Propag.*, vol. 52, no. 6, pp. 1568–1577, Jun. 2004.
14. L. Lu, G. Y. Li, A. L. Swindlehurst, A. Ashikhmin, and R. Zhang, "An overview of massive MIMO: Benefits and challenges," *IEEE J. Sel. Topics Signal Process.*, vol. 8, no. 5, pp. 742–758, Oct. 2014.
15. H. Q. Ngo, E. G. Larsson, and T. L. Marzetta, "Energy and spectral efficiency of very large multiuser MIMO systems," *IEEE Trans. Commun.*, vol. 61, no. 4, pp. 1436–1449, Apr. 2013.
16. M. R. Akdeniz, Y. Liu, M. K. Samimi, S. Sun, S. Rangan, T. S. Rappaport, and E. Erkip, "Millimeter wave channel modeling and cellular capacity evaluation," *IEEE J. Sel. Areas Commun.*, vol. 32, no. 6, pp. 1164–1179, Jun. 2014.
17. Z. Y. Jiang, A. F. Molisch, G. Caire, and Z. S. Niu, "Achievable rates of FDD massive MIMO systems with spatial channel correlation," *IEEE Trans. Wireless Commun.*, vol. 14, no. 5, pp. 2868–2882, May 2015.

Chapter 2
Background

From a signal processing point of view, the novel aspect of hybrid precoding/combining in comparison with the conventional fully digital precoding/combining lies in the introduction of the precoding/combining stage in the RF domain as a result of driving a large-scale antenna array by a limited number of RF chains. By treating the cascade of the RF stage and multiple-input multiple-output (MIMO) channel as the effective channel, the system model of massive MIMO is analogous to the conventional counterpart, where various solution techniques have been proposed for the single-user and multi-user scenarios. On the other hand, problem formulation and optimization of the RF component depend on the choice of the baseband component in a joint RF-baseband design. Therefore, before we embark on the study of hybrid precoding/combining design, we give a brief review of the related background and recent developments in this chapter, which serve as the basis for our proposed research in the subsequent chapters.

2.1 Linear Precoding and Combining for Point-to-Point MIMO

2.1.1 Baseband Digital Solutions for Conventional MIMO

Let us consider a point-to-point/single-user (SU) communication link, where a transmitter with N_t transmit antennas sends $N_\mathrm{s} = \min(N_\mathrm{t}, N_\mathrm{r})$ data streams to a receiver with N_r receive antennas. Specifically, the receive signal $\mathbf{y} \in \mathbb{C}^{N_\mathrm{r}}$ is given as

$$\mathbf{y} = \mathbf{H}\mathbf{x} + \mathbf{n} = \mathbf{H}\mathbf{F}\mathbf{s} + \mathbf{n},$$

T. Le-Ngoc, R. Mai, *Hybrid Massive MIMO Precoding in Cloud-RAN*, Wireless Networks, https://doi.org/10.1007/978-3-030-02158-0_2

where $\mathbf{H} \in \mathbb{C}^{N_r \times N_t}$ is the flat-fading MIMO channel matrix with (i, j)th entry representing the channel coefficient between the ith receive antenna and the jth transmit antenna, $\mathbf{x} = \mathbf{Fs} \in \mathbb{C}^{N_t}$ is the linearly precoded transmit signal with $\mathbf{F} \in \mathbb{C}^{N_t \times N_s}$ as the linear precoder and $\mathbf{s} \sim \mathcal{CN}(\mathbf{0}, \mathbf{I}_{N_s})$ as the Gaussian-coded data streams, and $\mathbf{n} \sim \mathcal{CN}(\mathbf{0}, \sigma_n^2 \mathbf{I}_{N_r})$ is zero-mean, spatially white Gaussian noise with variance σ_n^2. At the output of a linear receive combiner $\mathbf{W} \in \mathbb{C}^{N_r \times N_s}$, the signal estimate is given as

$$\hat{\mathbf{s}} = \mathbf{W}^{\mathrm{H}}\mathbf{HFs} + \mathbf{W}^{\mathrm{H}}\mathbf{n} = \mathbf{W}^{\mathrm{H}}\mathbf{U}\boldsymbol{\Lambda}\mathbf{V}^{\mathrm{H}}\mathbf{Fs} + \mathbf{W}^{\mathrm{H}}\mathbf{n},$$

where the second equality comes from the singular value decomposition (SVD) $\mathbf{H} = \mathbf{U}\boldsymbol{\Lambda}\mathbf{V}^{\mathrm{H}}$. From this relation, if the linear precoder $\mathbf{F}$ and linear combiner $\mathbf{W}$ are chosen as the N_s right and left singular vectors corresponding to the N_s dominant singular values of the channel $\mathbf{H}$, respectively, the effective channel $\mathbf{W}^{\mathrm{H}}\mathbf{HF}$ becomes diagonal. In other words, the Gaussian vector channel is converted into parallel non-interfering Gaussian scalar sub-channels. In combination with power allocation, it was shown in [1] that the channel capacity can be achieved. Equipped with this insight, joint design with respect to other criteria such as minimum mean square error (MMSE) and bit error rate (BER) minimization was also studied in [2–6]. In particular, with the definition of the mean square error (MSE) matrix as

$$\mathbf{E} = \mathbb{E}\left[\left(\hat{\mathbf{s}} - \mathbf{s}\right) \left(\hat{\mathbf{s}} - \mathbf{s}\right)^{\mathrm{H}} \right] = \mathbf{W}^{\mathrm{H}}\mathbf{HF}\mathbf{F}^{\mathrm{H}}\mathbf{H}^{\mathrm{H}}\mathbf{W} + \sigma_n^2 \mathbf{W}^{\mathrm{H}}\mathbf{W},$$

where the ith diagonal element of $\mathbf{E}$, i.e., $\mathbf{E}_{i,i}$, represents the MSE of the ith data stream s_i, the authors in [4] recognized that the popular design criteria such as mutual information, sum MSE, minimum BER, are in fact either Schur-convex or Schur-concave functions of the MSEs, i.e., diag($\mathbf{E}$), for which the optimal linear precoder must be chosen in the presence of a linear MMSE equalizer such that the effective channel $\mathbf{W}^{\mathrm{H}}\mathbf{HF}$ is diagonalized (up to a unitary rotation in the Schur-concave case). For both cases, the optimal linear precoder is derived from the right singular vectors of the MIMO channel $\mathbf{H}$ while the optimal linear MMSE combiner is the well-known linear Wiener filter, i.e.,

$$\mathbf{W}_{\mathrm{LMMSE}} = \left(\mathbf{HF}\mathbf{F}^{\mathrm{H}}\mathbf{H}^{\mathrm{H}} + \sigma_n^2 \mathbf{I}_{N_r}\right)^{-1} \mathbf{HF}.$$

In doing this, matrix optimization problems are reduced to convex vector optimization problems, which can be efficiently solved by waterfilling-based algorithms.

2.1.2 Progressive Baseband Digital Solutions for Conventional MIMO ARQ

In order to improve link reliability, modern wireless communications systems are often equipped with hybrid automatic repeat request (ARQ) mechanisms in practice.

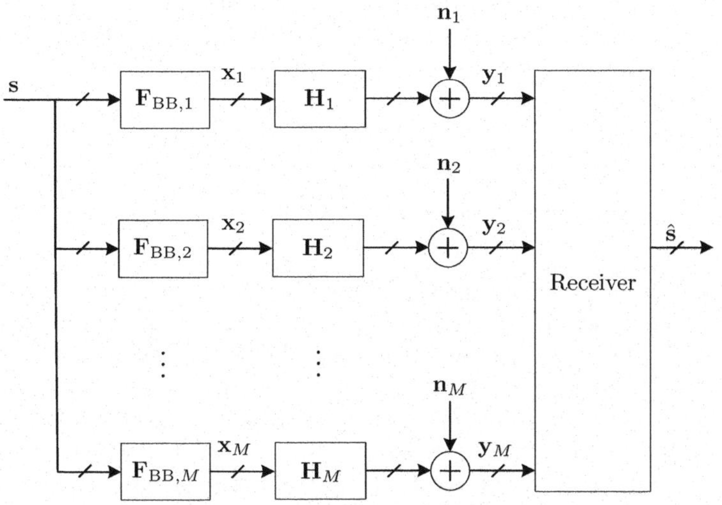

Fig. 2.1 Linear precoding and combining for M retransmissions of a packet

As mentioned in the previous section, linear MIMO precoding and combining are a useful technique to improve the efficiency and reliability of the system by exploiting the inherent spatial diversity created by multiple antennas. On the other hand, the combination of ARQ with precoding design makes it possible to exploit diversity from the temporal dimension. Such an idea is illustrated in Fig. 2.1, where each branch represents one round of ARQ retransmission, and the combiner performs linear combination across all retransmissions. In particular, the receive signal for the mth round of ARQ retransmission $\mathbf{y}_m \in \mathbb{C}^{N_\mathrm{r}}$ reads as

$$\mathbf{y}_m = \mathbf{H}_m \mathbf{F}_m \mathbf{s} + \mathbf{n}_m,$$

where $\mathbf{H}_m$ is the MIMO channel realization during the mth round of retransmission, $\mathbf{F}_m$ is the linear precoder, and $\mathbf{n}_m \sim \mathcal{CN}(\mathbf{0}, \sigma_n^2 \mathbf{I}_{N_\mathrm{r}})$ is the zero-mean, temporally and spatially white Gaussian noise with variance σ_n^2, i.e., $\mathbb{E}[\mathbf{n}_m \mathbf{n}_l^\mathrm{H}] = \delta_{ml}\sigma_n^2 \mathbf{I}_{N_\mathrm{r}}$. When viewed independently, each round of ARQ retransmission is nothing but a conventional point-to-point MIMO system as presented in the previous section. Following the observation that each retransmission is likely to experience independent channel fading, the current channel realization $\mathbf{H}_m$ is modeled as independent of the past realizations $\mathbf{H}_1, \ldots, \mathbf{H}_{m-1}$. If the linear precoder $\mathbf{F}_m$ is designed in accordance with the current channel realization $\mathbf{H}_m$ while taking into account past transmissions as represented by $\mathbf{F}_1, \ldots, \mathbf{F}_{m-1}$, this will allow us to leverage the temporal diversity and therefore further performance improvement can be expected. Towards this end, the objective function can be formulated as

$$\mathcal{I}(\mathbf{s}; \mathbf{y}_1, \mathbf{y}_2, \ldots, \mathbf{y}_m) = \log_2 \det \left(\mathbf{I}_{N_\mathrm{s}} + \sum_{i=1}^{m-1} \mathbf{F}_i^\mathrm{H} \mathbf{H}_i^\mathrm{H} \mathbf{H}_i \mathbf{F}_i + \mathbf{F}_m^\mathrm{H} \mathbf{H}_m^\mathrm{H} \mathbf{H}_m \mathbf{F}_m \right).$$

The unique aspect of MIMO ARQ is that past transmissions cannot be changed while future retransmissions might not be necessary. Therefore, it makes sense to optimize the precoder $\mathbf{F}_m$ only based on the current channel state information (CSI) $\mathbf{H}_m$ for each (re-)transmission while taking into account previous retransmissions $\sum_{i=1}^{m-1} \mathbf{F}_i^{\mathrm{H}}\mathbf{H}_i^{\mathrm{H}}\mathbf{H}_i\mathbf{F}_i$. Accordingly, the problem is formulated as

$$\sup_{\mathbf{F}_m} \mathcal{I}\left(\mathbf{s}; \mathbf{y}_1, \mathbf{y}_2, \ldots, \mathbf{y}_m\right) \quad \text{s.t. } \operatorname{tr}\left(\mathbf{F}_m\mathbf{F}_m^{\mathrm{H}}\right) \leq P_{\mathrm{t}}.$$

This is the basic idea behind sequential precoder design with respect to hybrid ARQ where the same data streams $\mathbf{s}$ are resent for each round of retransmission. The work in [7] solved such an optimization problem by showing that the optimal design must diagonalize the summation $\sum_{i=1}^{m} \mathbf{F}_i^{\mathrm{H}}\mathbf{H}_i^{\mathrm{H}}\mathbf{H}_i\mathbf{F}_i$. The issue of sum MSE minimization was addressed in [8]. Instead of the highly complex ML receiver as implied by the expression of $\mathcal{I}(\mathbf{s}; \mathbf{y}_1, \mathbf{y}_2, \ldots, \mathbf{y}_m)$, a linear precoder in conjunction with nonlinear MMSE-decision feedback equalizer (DFE) was derived in [9]. By relaxing the assumption of full CSI to partial CSI, i.e., channel transmit covariance, the objective of ergodic mutual information, i.e.,

$$\tilde{\mathcal{I}}\left(\mathbf{s}; \mathbf{y}_1, \mathbf{y}_2, \ldots, \mathbf{y}_m\right) = \mathbb{E}_{\mathbf{H}_m}\left[\log_2 \det\left(\mathbf{I}_{N_{\mathrm{s}}} + \sum_{i=1}^{m-1} \mathbf{F}_i^{\mathrm{H}}\mathbf{H}_i^{\mathrm{H}}\mathbf{H}_i\mathbf{F}_i + \mathbf{F}_m^{\mathrm{H}}\mathbf{H}_m^{\mathrm{H}}\mathbf{H}_m\mathbf{F}_m\right)\right]$$

is formulated and solved suboptimally in [10]. Extension to non-Gaussian inputs, i.e., practical modulation schemes, for mutual information maximization was made in [11].

2.1.3 Hybrid RF-Baseband Solutions for Massive MIMO

Let us shift our focus to a narrowband massive MIMO system as illustrated in Fig. 2.2, where the number of transmit antennas N_{t} and that of receive antennas N_{r} are on the order of tens or even hundreds, i.e., $N_{\mathrm{t}}, N_{\mathrm{r}} \gg 1$. As explained in Sect. 1.2.1, because of high hardware complexity and power consumption, the fully digital architecture is not practically realizable for massive MIMO. Instead, the large-scale antenna arrays are driven by a limited number of n_{t} RF chains at the transmitter and n_{r} RF chains at the receiver with $n_{\mathrm{t}} \ll N_{\mathrm{t}}$ and $n_{\mathrm{r}} \ll N_{\mathrm{r}}$. In this case, the number of data streams that can be physically supported must satisfy $N_{\mathrm{s}} \leq \min(n_{\mathrm{t}}, n_{\mathrm{r}})$. The received signal reads as

$$\mathbf{y} = \mathbf{H}\mathbf{x} + \mathbf{n},$$

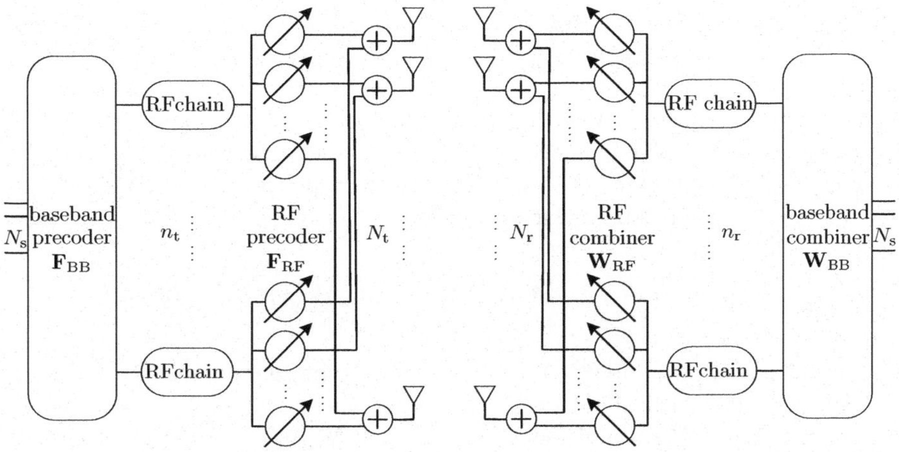

Fig. 2.2 Hybrid precoding and combining with RF phase-shifters

where $\mathbf{H} \in \mathbb{C}^{N_r \times N_t}$ is the block fading channel matrix normalized as $\mathbb{E}[||\mathbf{H}||_F^2] = N_t N_r$, $\mathbf{x} \in \mathbb{C}^{N_t}$ is the transmitted signal satisfying the normalized power constraint $\mathrm{tr}(\mathbb{E}[\mathbf{x}\mathbf{x}^H]) \le P_t$, and $\mathbf{n} \sim \mathcal{CN}(\mathbf{0}, \sigma_n^2 \mathbf{I}_{N_r})$ is the zero-mean, spatially white Gaussian noise with variance σ_n^2. In particular,

$$\mathbf{x} = \mathbf{F}\mathbf{s} = \mathbf{F}_{RF}\mathbf{F}_{BB}\mathbf{s},$$

where $\mathbf{F} = \mathbf{F}_{RF}\mathbf{F}_{BB}$ is a linear precoder composed of two stages: an RF analog component $\mathbf{F}_{RF} \in \mathbb{C}^{N_t \times n_t}$ and a baseband digital component $\mathbf{F}_{BB} \in \mathbb{C}^{n_t \times N_s}$, and $\mathbf{s} \sim \mathcal{CN}(\mathbf{0}, \mathbf{I}_{N_s})$ is the normalized data stream. The signal estimate is given by

$$\hat{\mathbf{s}} = \mathbf{W}^H \mathbf{y} = \mathbf{W}_{BB}^H \mathbf{W}_{RF}^H \mathbf{y},$$

where $\mathbf{W} = \mathbf{W}_{RF}\mathbf{W}_{BB}$ is a linear two-stage combiner consisting of an RF analog combiner $\mathbf{W}_{RF} \in \mathbb{C}^{N_r \times n_r}$ and a baseband digital combiner $\mathbf{W}_{BB} \in \mathbb{C}^{n_r \times N_s}$. To reduce hardware complexity, RF elements might consist of only phase-shifters. In other words, the elements of the RF precoder $\mathbf{F}_{RF}$ and RF combiner $\mathbf{W}_{RF}$ are of constant modulus, i.e., $|[\mathbf{F}_{RF}]_{i,j}| = \frac{1}{\sqrt{N_t}}$, $i = 1, \ldots, N_t$, $j = 1, \ldots, n_t$ and $|[\mathbf{W}_{RF}]_{p,q}| = \frac{1}{\sqrt{N_r}}$, $p = 1, \ldots, N_r$, $q = 1, \ldots, n_r$. On the assumption of perfect CSI, one noticeable difference in the formulation of hybrid precoding and combining from its fully digital counterpart in Sect. 2.1.1 is the introduction of the RF stages $\mathbf{F}_{RF}$ and $\mathbf{W}_{RF}$. Although by treating $\mathbf{W}_{RF}^H \mathbf{H} \mathbf{F}_{RF}$ as the effective MIMO channel, previous baseband digital solutions for conventional MIMO are readily applicable, the RF components nonetheless need to be carefully optimized for the best performance. For example, optimal power loading schemes for fully digital solutions are unlikely to remain optimal without taking into account the presence of the RF precoder since $\mathrm{tr}(\mathbb{E}[\mathbf{x}\mathbf{x}^{\mathrm{H}}]) = \|\mathbf{F}_{\mathrm{RF}}\mathbf{F}_{\mathrm{BB}}\|_{\mathrm{F}}^2 \leq P_{\mathrm{t}}$. In combination with the nonconvex constant-modulus constraints, such problems are in fact challenging to solve.

To jointly optimize the RF precoder and combiner, one may attempt to maximize the capacity of the effective channel. This motivates the consideration of the mutual information as the criterion, which is defined between the baseband-precoded signal and the signal estimate after linear RF combining, i.e.,

$$\mathcal{I}_{\mathrm{RF}} = \log_2 \det\left(\mathbf{I}_{N_{\mathrm{s}}} + \frac{P_{\mathrm{t}}}{N_{\mathrm{s}}}\mathbf{F}_{\mathrm{RF}}^{\mathrm{H}}\mathbf{H}^{\mathrm{H}}\mathbf{W}_{\mathrm{RF}}\left(\mathbf{W}_{\mathrm{RF}}^{\mathrm{H}}\mathbf{W}_{\mathrm{RF}}\right)^{-1}\mathbf{W}_{\mathrm{RF}}^{\mathrm{H}}\mathbf{H}\mathbf{F}_{\mathrm{RF}}\right),$$

with equal power allocation. It was shown in [12] that the optimal RF precoder (combiner) is related to the right (left) singular vectors corresponding to the n_{t} (n_{r}) largest singular values of the MIMO channel $\mathbf{H}$. Built upon this insight, the constant-modulus RF precoder and combiner were further derived by extracting the phases of their optimal counterparts. In doing this, the solutions can adapt flexibly to the instantaneous channel variation. The downside is that the optimal solutions need to be computed first based on the perfect CSI. Such requirements were relaxed in [13] where the hybrid combining aspect was addressed with respect to the channel receive correlation. On the other hand, when strong spatial correlation is present in the MIMO channel, one simplifying and effective alternative is to reduce the design of RF transmit and receive phase-shifting to beam selection from the N_{t}-dimensional discrete Fourier transform (DFT) codebook $\mathcal{D}_{\mathrm{t}} = [\frac{1}{\sqrt{N_{\mathrm{t}}}}e^{-\frac{\jmath 2\pi mn}{N_{\mathrm{t}}}}]$, $m, n = 0, \ldots, N_{\mathrm{t}} - 1$ and the N_{r}-dimensional DFT codebook $\mathcal{D}_{\mathrm{r}} = [\frac{1}{\sqrt{N_{\mathrm{r}}}}e^{-\frac{\jmath 2\pi pq}{N_{\mathrm{r}}}}]$, $p, q = 0, \ldots, N_{\mathrm{r}} - 1$, respectively [14]. Different from the independently and identically distributed (i.i.d.) Rayleigh fading channel which corresponds to a rich scattering environment, a strongly correlated channel experiences limited scattering, and thus has a directional power spectrum in the angle domain. Intuitively, if the chosen DFT beams are well aligned with the directional scattering, it is unlikely for such an RF beamformer to cause a severe loss of channel power. Nonetheless, since the DFT beamforming directions correspond to uniformly quantized angles of arrival (AoAs)/angles of departure (AoDs), phase-shifting adaptive to instantaneous CSI as in [12] is more advantageous as the channel becomes increasingly independent. Besides, in a multi-user setting, the coarse resolution provided by DFT might not be sufficient to achieve good spatial separation between users or good coverage when the users are not located in a favorable direction.

The work in [15] treated perfect CSI-based hybrid precoding and combining with RF phase-shifters in the context of mmWave massive MIMO systems. Subject to the power constraint $\|\mathbf{F}_{\mathrm{RF}}\mathbf{F}_{\mathrm{BB}}\|_{\mathrm{F}}^2 \leq P_{\mathrm{t}}$, the hybrid precoder and combiner are obtained by optimizing the mutual information $\mathcal{I}(\mathbf{s}; \mathbf{y})$, i.e.,

$$\mathcal{I}(\mathbf{s}; \mathbf{y}) = \log_2 \det\left(\mathbf{I}_{N_{\mathrm{s}}} + \frac{P_{\mathrm{t}}}{N_{\mathrm{s}}}\mathbf{R}_n^{-1}\mathbf{W}_{\mathrm{BB}}^{\mathrm{H}}\mathbf{W}_{\mathrm{RF}}^{\mathrm{H}}\mathbf{H}\mathbf{F}_{\mathrm{RF}}\mathbf{F}_{\mathrm{BB}}\mathbf{F}_{\mathrm{BB}}^{\mathrm{H}}\mathbf{F}_{\mathrm{RF}}^{\mathrm{H}}\mathbf{H}^{\mathrm{H}}\mathbf{W}_{\mathrm{RF}}\mathbf{W}_{\mathrm{BB}}\right)$$

where $\mathbf{R}_n \triangleq \sigma_n^2 \mathbf{W}_{\mathrm{BB}}^{\mathrm{H}} \mathbf{W}_{\mathrm{RF}}^{\mathrm{H}} \mathbf{W}_{\mathrm{RF}} \mathbf{W}_{\mathrm{BB}}$ is the noise correlation after hybrid combining. In contrast with the study in [12] where the RF precoder and combiner were optimized with respect to the RF-processed channel capacity, the effect of baseband precoding and combining are explicitly incorporated in seeking a joint RF-baseband design. Due to the difficulty posed by the nonconvex constant-modulus constraint on the RF elements, joint optimization of the hybrid precoder $\{\mathbf{F}_{\mathrm{RF}}, \mathbf{F}_{\mathrm{BB}}\}$ and hybrid combiner $\{\mathbf{W}_{\mathrm{RF}}, \mathbf{W}_{\mathrm{BB}}\}$ is not possible. Instead, the authors focused on joint optimization of RF-baseband precoding (combining) by abstracting the effect of hybrid combining (precoding). In view of the constant-modulus property of the array steering vectors, the design of RF precoder (RF combiner) was formulated as selecting the optimal array steering vectors corresponding to all path AoAs (AoDs). In combination with the baseband digital precoder (combiner), such a hybrid design was cast as matrix reconstruction of the fully digital optimal solution for conventional MIMO as in Sect. 2.1.1, and a near-optimal precoding (combining) solution was found via an orthogonal matching pursuit algorithm. Although low-complexity, one immediate drawback of this technique is that extensive information of AoAs (AoDs) for all the propagation paths is required. In an attempt to address this practical limitation, the idea of sparse approximation was further applied to channel estimation in [16]. The sub-optimal restriction of RF elements to the space of array steering vectors incurs severe performance loss under certain channel conditions, and therefore, general formulation of constant-modulus RF design is needed. Following the idea of matrix reconstruction, improved hybrid solutions were devised in [17] using Grassmann manifold optimization and in [18] based on approximate joint diagonalization of matrices. To obviate the need for computing the optimal solution, a heuristic method was proposed in [19]. In [20], multi-layer codebook design for both RF phase-shifters and baseband precoder was considered for wideband orthogonal frequency-division multiplexing (OFDM) systems. In this case, beam-steering in the RF domain is commonly applied across all sub-carriers while baseband precoding is specific to each sub-carrier.

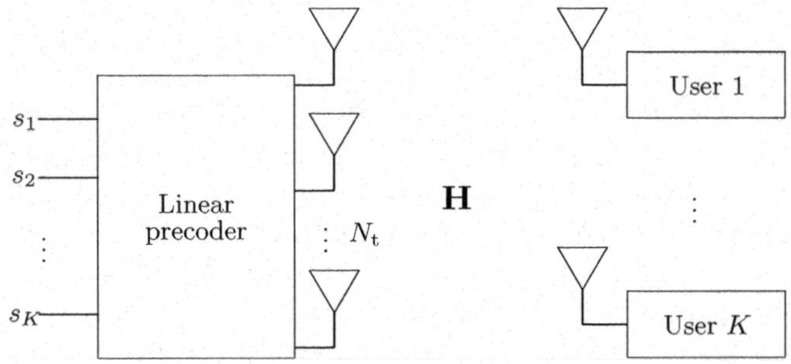

Fig. 2.3 MU-MIMO downlink linear precoding for single-antenna users

2.2 Multi-User MIMO Precoding

2.2.1 *Precoding for Conventional MIMO: From Linear to Nonlinear*

As illustrated in Fig. 2.3, let us consider a traditional multi-user (MU)-MIMO broadcast channel (BC) where a base station (BS) with N_t transmit antennas is serving $K \leq N_t$ single-antenna users. Let $\mathbf{h}_i^H \in \mathbb{C}^{1\times N_t}$ represent the channel between the ith user and the BS, and $\mathbf{H} \triangleq [\mathbf{h}_1, \ldots, \mathbf{h}_K]^H \in \mathbb{C}^{K\times N_t}$. The received signals for all K users can be collectively written as

$$\mathbf{y} = [\mathbf{h}_1, \ldots, \mathbf{h}_K]^H \mathbf{x} + \mathbf{n} = \mathbf{H}\mathbf{x} + \mathbf{n},$$

where $\mathbf{x} \in \mathbb{C}^{N_t}$ is the transmit signal, and $\mathbf{n} \sim \mathcal{CN}(\mathbf{0}, \sigma_n^2 \mathbf{I}_K)$ is the zero-mean, spatially white Gaussian noise with variance σ_n^2. It is well-known that the sum capacity is achievable by encoding $\mathbf{x}$ by dirty paper coding (DPC) [21, 22]. However, DPC involves sophisticated random coding and binning approaches, and therefore is too computationally intensive to be useful in practice. Alternatively, a linear precoding scheme, channel inversion or zero forcing (ZF), is usually preferred because of its low complexity and being not too sensitive to channel estimation errors. At the expense of transmit power enhancement, inter-user interference is completely eliminated. In this case, the transmit signal is given as

$$\mathbf{x} = \beta^{-1}\mathbf{H}^H \left(\mathbf{H}\mathbf{H}^H\right)^{-1} \mathbf{s},$$

where β^{-1} is the power scaling factor to enforce the power constraint $\text{tr}(\mathbb{E}[\mathbf{x}\mathbf{x}^H]) \leq P_t$, and $\mathbf{s} = [s_1, s_2, \ldots, s_K]^T$ represents the data streams for the K users. When multi-user diversity is present in the system, i.e., there is a large number of users with heterogeneous channel realizations, it is possible to achieve the sum channel capacity by simply combining ZF precoding with greedy user selection [23, 24]. However, the effectiveness of such linear schemes might become questionable when users experience the same large-scale channel fading and require an equal-rate performance. If the system is fully loaded $\frac{N_t}{K} = 1$, i.e., the number of spatially multiplexed users K is equal to the maximum number of data streams N_t that can be physically supported, the sum rate unfortunately does not grow linearly with the number of users [25]. Note that the MIMO channel matrix $\mathbf{H} \in \mathbb{C}^{K\times N_t}$ in this case is square. It was observed in [25] that all the eigenvalues except one of $\mathbf{H}$ have magnitudes of comparable order. On the contrary, the inverse of the peculiar, or ill-behaving, eigenvalue has an infinite mean. Intuitively, this implies that when zero-forcing (channel inversion) precoding is used to serve the K homogeneous users, transmission along the eigenvector associated with the ill-behaving eigenvalue would consume most of the power. As a result, the sum rate performance does not grow linearly with the number of users.

To avoid transmission on the sub-channel with a poor channel gain, one possible solution is to use regularized ZF (RZF) precoding [25], where instead of direct inversion of the channel, a regularized channel is inverted. In other words,

$$\mathbf{x} = \beta^{-1}\mathbf{H}^{\mathrm{H}}\left(\mathbf{H}\mathbf{H}^{\mathrm{H}} + \alpha\mathbf{I}\right)^{-1}\mathbf{s}, \tag{2.1}$$

where α is the regularization factor to be optimized. Clearly, such an RZF precoding solution includes channel inversion as a special case when $\alpha = 0$. In other words, the introduction of the regularization factor α represents an additional degree of freedom (DoF) for beamforming optimization. Under the assumption that users receive the same signal-to-interference-plus-noise ratio (SINR), it was proved in [25] that the asymptotically optimal $\alpha^\star$ in the sense of SINR maximization is $\alpha^\star = \frac{K\sigma_n^2}{P_\mathrm{t}}$. Such an RZF precoder can be equivalently obtained by an MMSE formulation[26]. In addition to modifying the linear precoding front-end, modification of the signal structure is another feasible avenue. Here, as illustrated in Fig. 2.4, in lieu of the original data symbol $\mathbf{d}$, its perturbed version $\mathbf{s}$ is transmitted [27], i.e.,

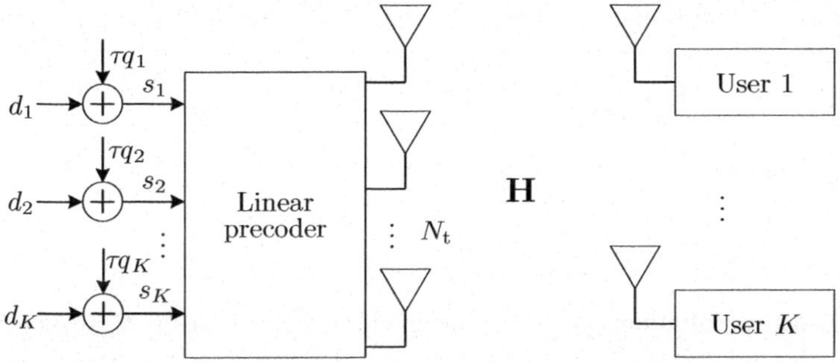

Fig. 2.4 MU-MIMO downlink nonlinear precoding for single-antenna users

$$\mathbf{s} = \mathbf{d} + \tau\mathbf{q}, \tag{2.2}$$

where $\mathbf{q} \in \mathbb{Z}^K + \jmath\mathbb{Z}^K$ is a complex integer vector, and $\tau > 0$ is a design parameter chosen to ensure correct recovery of $\mathbf{s}$ by modulo decoding. In particular, by defining the element-wise modulo-τ operation for a real vector $\mathbf{z}$ as

$$f_\tau(\mathbf{z}) = \mathbf{z} - \left\lfloor \frac{\mathbf{z}}{\tau} + 0.5 \right\rfloor \tau,$$

it is required that

$$f_\tau(\Re\{\mathbf{s}\}) + \jmath f_\tau(\Im\{\mathbf{s}\}) = \mathbf{d}.$$

In other words, enhancement of the system performance is enabled through introducing extra DoFs in the signal domain, as represented by the addition and removal of a perturbation vector $\mathbf{q}$. To search for $\mathbf{q}$, the authors in [27] proposed to minimize the transmit power as the cost function, or equivalently,

$$\mathbf{q}_{\text{pow}} = \underset{\mathbf{q}\in\mathbb{Z}^K+\jmath\mathbb{Z}^K}{\arg\min} \left\|\mathbf{H}^{\text{H}}(\mathbf{H}\mathbf{H}^{\text{H}} + \xi\mathbf{I})^{-1}\,(\mathbf{d}+\tau\mathbf{q})\,\right\|^2,$$

where $\xi = 0$ and $\xi = \alpha^{\star}$ correspond to using ZF and RZF precoding as the linear front-end, respectively. Because of the nonlinear nature of the search for $\mathbf{q}$, the probability distribution of $\mathbf{s}$ in (2.2) is generally not known. In an attempt to circumvent this difficulty, the MSE in [28] is defined as the squared distance between the perturbed signal and the signal estimate before modulo decoding conditioned on $\mathbf{d}$, i.e.,

$$\text{MSE}|_{\mathbf{d}} = \mathbb{E}\left[\|\mathbf{s}-\beta\mathbf{y}\|^2\,|\mathbf{d}\right],$$

where the averaging is over noise. Intuitively, if the signal estimate $\beta\mathbf{y}$ is close to the perturbed signal $\mathbf{s}$, the recovered data streams should be close to the original data streams $\mathbf{d}$ after the removal of the perturbation vector by modulo decoding. Minimization of $\text{MSE}|_{\mathbf{d}}$ results in the RZF precoder in (2.1) with the same $\alpha^{\star} = \frac{K\sigma_n^2}{P_t}$, and the perturbation vector

$$\mathbf{q}_{\text{mse}} = \underset{\mathbf{q}\in\mathbb{Z}^K+\jmath\mathbb{Z}^K}{\arg\min} \|-\mathbf{L}\mathbf{d}-\tau\mathbf{L}\mathbf{q}\|^2,$$

where the lower triangular matrix $\mathbf{L}$ is defined through Cholesky factorization, i.e.,

$$\left(\mathbf{H}\mathbf{H}^{\text{H}} + \alpha^{\star}\mathbf{I}\right)^{-1} = \mathbf{L}^{\text{H}}\mathbf{L}.$$

In terms of coded BER, the MSE-based solution yields a noticeably superior performance than the power minimization-based counterpart [28].

Since the perturbation vector is a Gaussian integer, the search can be viewed as a problem of closest point search in a lattice. If $\mathbf{s}$ is picked from a high-order square constellation such as 16-, 64-, and 256-QAM as defined in Long-Term Evolution (LTE), and the optimal perturbation vector is optimally found, then by approximating the equiprobable discrete constellation points as continuous and uniformly distributed in a hyper-rectangle, the resulting errors from the search can be accordingly treated as uniformly distributed [29]. If the perturbation vector is obtained by power minimization, the error is nothing but the transmit power. In contrast to [30] where the transmit power is obtained by numerically solving a set of fixed-point equations, such lattice-theoretic approximation yields a mathematically tractable lower bound. This insight was leveraged for channel vector quantization design in [31], and for greedy user selection to alleviate the concern of power enhancement in cooperative ZF beamforming [32]. While the per-BS power constraint is more practical for multi-cell downlink beamforming, unfortunately, the MMSE-VP precoder has to be numerically optimized [33].

When users are equipped with multiple antennas, an enhanced system performance is expected from joint design of nonlinear precoding and linear combining. So far, the focus has been exclusively on non-iterative methods for the benefit of low complexity. The basic idea is to first use block diagonalization (BD) to eliminate inter-user interference and hence create parallel SU-MIMO channels, and then perform VP across the spatially multiplexed data streams in conjunction with SU-MIMO precoding and combining for each user. In [34], over each SU-MIMO channel, ZF-VP is used for spatial multiplexing while treating each receive antenna as a virtual user. In doing so, users only need to know the power scaling factor β for detection. Such a design, motivated by the need for signaling overhead reduction, was shown to approach the performance of waterfilling-based solutions [1]. The work in [35] instead combined BD with MMSE-VP, and demonstrated the benefit of geometric mean decomposition-based joint design [5] in terms of improved BER. By exploiting uniform channel decomposition [6], the linear MMSE receiver could also be incorporated [36, 37], which further decreases the BER. Assuming matched filtering at the users, a non-iterative approach for cooperative ZF-VP beamforming was proposed in [38] and compared with the linear counterpart.

2.2.2 *Hybrid RF-Baseband Solutions for Massive MIMO*

For a multi-user massive MIMO system where the BS employs the hybrid precoding architecture, the transmit signal is expressed as

$$\mathbf{x} = \mathbf{F}\mathbf{s} = \mathbf{F}_{\mathrm{RF}}\mathbf{F}_{\mathrm{BB}}\mathbf{s}.$$

Subject to the transmit power constraint $\|\mathbf{F}_{\mathrm{RF}}\mathbf{F}_{\mathrm{BB}}\|_{\mathrm{F}}^2 \leq P_{\mathrm{t}}$, joint RF-baseband optimization can be performed sequentially in light of the following equivalence [39]:

$$\inf_{\{\mathbf{F}_{\mathrm{RF}},\mathbf{F}_{\mathrm{BB}}\}} \phi\left(\mathbf{F}_{\mathrm{RF}}, \mathbf{F}_{\mathrm{BB}}\right) = \inf_{\mathbf{F}_{\mathrm{RF}}} \left\{ \inf_{\mathbf{F}_{\mathrm{BB}}} \phi\left(\mathbf{F}_{\mathrm{RF}}, \mathbf{F}_{\mathrm{BB}}\right) \right\} = \inf_{\mathbf{F}_{\mathrm{BB}}} \left\{ \inf_{\mathbf{F}_{\mathrm{RF}}} \phi\left(\mathbf{F}_{\mathrm{RF}}, \mathbf{F}_{\mathrm{BB}}\right) \right\},$$

which holds for *any* cost function $\phi\left(\cdot\right)$. Here, the digital precoder $\mathbf{F}_{\mathrm{BB}}$ at baseband generally employs traditional MU-MIMO linear schemes such as ZF and RZF, and the interest lies mainly in deriving the RF analog precoder $\mathbf{F}_{\mathrm{RF}}$ subject to additional constraints such as constant-modulus elements, i.e., phase-shifters, and partial CSI.

RF phase-shifting design with perfect CSI was addressed in [40, 41]. In [40], the phase-shifters in the RF domain are heuristically derived based on extracting the phases of the maximum ratio transmission (MRT) beamformer. In combination with

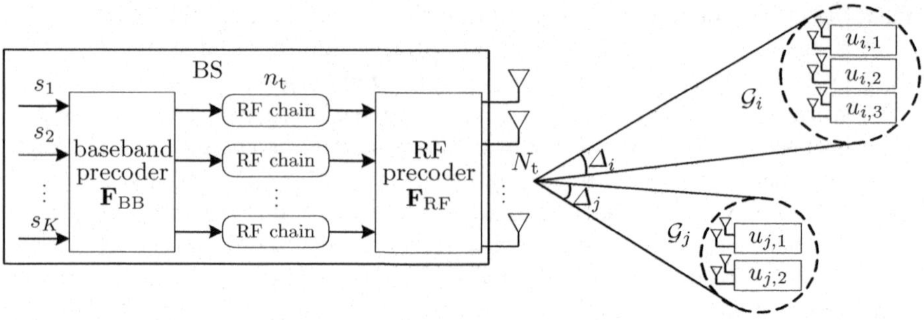

Fig. 2.5 Massive MIMO downlink precoding for clustered users

the baseband ZF precoding, the hybrid solution was shown to perform close to the fully digital ZF counterpart in terms of sum rate. This idea was further pursued in a multi-antenna user setting in [41], where hybrid combining was also considered with the constant-modulus RF combiner reduced to DFT beam selection.

It is worth mentioning that one basic assumption in the aforementioned works [15–19, 40, 41] is that perfect knowledge of the high-dimensional MIMO channel is available at the BS. This assumption, however, could be problematic in closed-loop frequency-division duplexing (FDD) systems, in that an enormous amount of channel estimates need to be frequently fed back [42]. In an effort to remedy the issue with channel estimation overhead, one interesting idea is to adjust RF processing solely based on the statistical CSI while updating baseband processing according to the instantaneous effective CSI [43–47]. The slow-varying nature of statistical CSI renders it unnecessary for frequent update which leads to reduction in feedback overhead. In the presence of the RF stage, the dimension of the effective channel from the perspective of the baseband is significantly decreased in contrast to the original MIMO channel, and thus timely update of instantaneous CSI becomes feasible.

So far, studies on two-timescale hybrid precoding have been largely focused on scenarios where single-antenna users are assumed to be clustered. According to the one-ring correlation channel model [48, 49], the transmit correlation matrix is related to the mean AoA and angle spread, which are in turn decided by a user's relative location to the BS and its surrounding scattering environment. Therefore, it makes sense to geographically divide users into clusters, and assume that users associated with the same cluster share the same transmit correlation matrix, as illustrated in Fig. 2.5. This observation motivates us to view the interference experienced by each user as a combination of *inter*-cluster and *intra*-cluster interference, which can be handled by RF precoding and baseband precoding, respectively. The authors in [43, 44] proposed the technique of joint spatial division and multiplexing where the statistical CSI-based RF precoder is employed to separate users into non-interfering clusters by BD. This technique relies on the premise that the sub-spaces spanned by the transmit correlation matrices do not significantly overlap with each other.

Depending on the availability of CSI, baseband precoding can be carried out jointly across all user clusters or separately for each cluster of users. It was concluded that the sum rate of per-cluster precoding would become interference limited when inter-cluster interference is not effectively suppressed. Instead of trying to null inter-cluster interference as in BD, the work in [50] derived the RF precoder with the aim of striking a balance between cluster-wise self-transmission and interference leakage power. A lower bound for the cost function, the expected signal-to-leakage-plus-noise ratio (SLNR), was proposed and solved by well-known trace quotient algorithms. Not surprisingly, since self-transmission is not overly penalized by allowing controlled inter-cluster interference, the sum rate is increased. The authors in [46] considered a design of statistical CSI-based RF phase-shifting to maximize the worst ergodic rate. Based on the insight that the DFT matrix well approximates the transmit correlation matrix for uniform linear arrays (ULAs) in the large array regime ($N_\mathrm{t} \rightarrow \infty$), the problem is reduced to DFT beam selection. When taking into account channel estimation errors, such constant-modulus solutions are even able to outperform the fully digital ZF solution. In [51], the MSE across all users is minimized for OFDM systems where elements in both the RF and the baseband domain are constrained to be of constant modulus in an attempt to avoid high peak-to-average power ratio.

It is worth mentioning that the effectiveness of the hybrid precoding solutions with two-timescale CSI as in [43, 50, 52] relies on the assumption that users are naturally partitioned into groups, and the same group of users experiences identical channel spatial correlation. This is however too restrictive in practice. Thus, the work in [53–55] proposed various user grouping algorithms based on the distance between the subspaces of the transmit spatial correlation, and RF beamforming was adapted to the centroid of the correlation matrices for each user group. Since the mean correlation is only a rough approximation, in this case, it is unlikely for the resulting RF beams to create near-perfect spatial separation, which casts doubt on the feasibility of group-wise spatial multiplexing at baseband.

When it comes to multi-cell systems, additional design issues, such as limited inter-BS cooperation in terms of signal and local CSI exchange, per-BS power constraint, and inter-cell interference need to be properly addressed. For example, in [45], only statistical CSI was assumed to be globally available, and the cluster-wise RF precoding was constrained to the null space of the superimposed transmit correlation matrices for interference reduction. In conjunction with local CSI-based ZF solutions, the RF solutions are derived to maximize a general utility function of spectral efficiency. In [52], RF precoding is designed with the objective of minimizing interference leakage power with linear pricing. Under the assumption that statistical CSI used for RF precoder update is outdated, a subspace tracking and compensation algorithm on Grassmann manifolds was proposed. The concept of deterministic equivalent was exploited in [47] to approximate the SINR chance constraint by deterministic functions, where accordingly, the RF precoding solutions are obtained. In [56], MMSE-VP was employed for instantaneous CSI-based two-stage precoder design in cooperative multi-cell systems.

2.3 Summary

For point-to-point massive MIMO systems, previous work has shown that even with a reduced number of RF chains, instantaneous CSI-based two-stage precoding/combining solutions are capable of delivering a performance comparable to their fully digital counterparts. This is especially the case when the MIMO channel is correlated, as commonly found in a directional propagation environment such as mmWave channels. However, because of the tremendous overhead of estimating the high-dimensional MIMO channel, one might be interested to know if such an observation is still valid when perfect CSI in the RF domain is relaxed to statistical CSI, e.g., channel covariance. Besides, it is noted that joint precoder-combiner optimization in conventional MIMO gives a substantial performance enhancement. Nonetheless, such an approach has not been attempted to derive the hybrid precoding and combining solutions yet. Finally, the design and evaluation of statistical CSI-based RF phase-shifting remain an open issue.

When packet retransmission is incorporated through hybrid ARQ mechanisms in conventional MIMO systems, previous research efforts have demonstrated that by exploiting the temporal diversity in the linear precoder optimization, the system performance can be improved. Unfortunately, such solutions cannot be directly applied to massive MIMO. To begin with, it has been assumed that the received signals from the past rounds of retransmission are fully accessible to the baseband. However, such an assumption raises concern about the storage requirement and processing complexity when the received signals are of high dimensions. A potential remedy is to introduce hybrid RF-baseband combining which reduces the dimension of the received signals to be combined through RF preprocessing. In doing this, the baseband has only access to the low-dimensional received signals at the output of the RF combiner. On the other hand, when the hybrid precoding structure with RF phase-shifting is employed, it is necessary for the optimization of baseband precoding to take into account the design of RF phase-shifting and hybrid combining at the receiver. Hence, a novel design of hybrid precoding and combining is required.

In a multi-user massive MIMO environment, the principle of hybrid RF-baseband precoding is to first create spatial separation of users in the RF beam domain and then perform spatial multiplexing at baseband. In particular, by assuming that users are geographically clustered in hotspots, it suffices to adjust the RF precoder based on the statistical CSI in consideration of the baseband precoder adaptive to the instantaneous effective CSI. As a result, only two-timescale CSI is required, which significantly reduces the channel estimation overhead. It is worth mentioning that the existing work has restricted the attention to linear precoding schemes such as ZF and RZF at baseband. Unfortunately, the linear schemes suffer severe power loss when a maximum number of equal-rate users is spatially multiplexed. On the contrary, by introducing a perturbation vector as additional DoFs for performance optimization, VP effectively addresses such an issue. A direct consequence of hybrid precoding with a reduced number of RF chains is that the number of data streams that can be physically supported is limited. Thus, it is desirable for hybrid precoding to perform well in fully loaded systems for utility maximization. In view of the drawback of linear schemes in this case, it is natural to explore how nonlinear VP techniques can be combined with the two-timescale CSI-based hybrid precoding design.

When extended to a multi-cell massive MIMO environment, the hybrid precoder needs to take into account the presence of inter-cell interference as well. The effectiveness of inter-cell interference mitigation depends on the degree of inter-cell cooperation in terms of information exchange allowed. For network MIMO processing, the technique of hybrid precoding design for single cell cannot directly carry over. In particular, because of the per-BS power constraint, closed-form expressions for the linear baseband front-end are no longer available. Given the difficulty with optimizing the statistical CSI-based RF precoder with respect to the traditional performance metrics, e.g., mutual information and MSE, the existing work has largely relied on heuristics. We remark that although the use of two-timescale CSI provides an effective alternative to addressing the issue of channel estimation overhead, it is somewhat restrictive from the perspectives of optimization and applicability. For example, in the absence of the analytical baseband solutions, iterative procedures are generally required for joint RF-baseband optimization. It is not clear how such alternating optimization can be carried out when the design variables are adaptive to different time scales. Besides, the assumption that users are geographically clustered and the user clusters are separated apart can turn out to be too ideal. Hence, novel design approaches that overcome such disadvantages while enjoying comparable channel estimation overhead with the two-timescale CSI are desired.

References

1. E. Telatar, "Capacity of multi-antenna Gaussian channels," *European Trans. Telecommun.*, vol. 10, no. 6, pp. 585–595, Nov. 1999.
2. H. Sampath, P. Stoica, and A. Paulraj, "Generalized linear precoder and decoder design for MIMO channels using the weighted MMSE criterion," *IEEE Trans. Commun.*, vol. 49, no. 12, pp. 2198–2206, Dec. 2001.
3. A. Scaglione, P. Stoica, S. Barbarossa, G. B. Giannakis, and H. Sampath, "Optimal designs for space-time linear precoders and decoders," *IEEE Trans. Signal Process.*, vol. 50, no. 5, pp. 1051–1064, May 2002.
4. D. P. Palomar, J. M. Cioffi, and M. A. Lagunas, "Joint Tx-Rx beamforming design for multicarrier MIMO channels: A unified framework for convex optimization," *IEEE Trans. Signal Process.*, vol. 51, no. 9, pp. 2381–2401, Sep. 2003.
5. Y. Jiang, J. Li, and W. W. Hager, "Joint transceiver design for MIMO communications using geometric mean decomposition," *IEEE Trans. Signal Process.*, vol. 53, no. 10, pp. 3791–3803, Oct. 2005.
6. ——, "Uniform channel decomposition for MIMO communications," *IEEE Trans. Signal Process.*, vol. 53, no. 11, pp. 4283–4294, Nov. 2005.
7. H. Sun, H. Samra, Z. Ding, and J. Manton, "Constrained capacity of linear precoded ARQ in MIMO wireless systems," in *Proc. IEEE Int. Conf. Acoustics, Speech and Signal Process.*, Philadelphia, PA, USA, Mar. 2005.

8. H. Sun, J. H. Manton, and Z. Ding, "Progressive linear precoder optimization for MIMO packet retransmissions," *IEEE J. Sel. Areas Commun.*, vol. 24, no. 5, pp. 448–456, Mar. 2006.
9. H. Sun and Z. Ding, "Iterative transceiver design for MIMO ARQ retransmissions with decision feedback detection," *IEEE Trans. Signal Process.*, vol. 55, no. 7, pp. 3405–3416, Jul. 2007.
10. H. Sun, Z. H. Shi, C. M. Zhao, J. H. Manton, and Z. Ding, "Progressive linear precoder optimization for MIMO packet retransmissions exploiting channel covariance information," *IEEE Trans. Commun.*, vol. 56, no. 5, pp. 818–827, May 2008.
11. X. Liang, C. M. Zhao, and Z. Ding, "Sequential linear MIMO precoder optimization for hybrid ARQ retransmission of QAM signals," *IEEE Commun. Lett.*, vol. 15, no. 9, pp. 913–915, Sep. 2011.
12. X. Zhang, A. F. Molisch, and S.-Y. Kung, "Variable-phase-shift-based RF-baseband codesign for MIMO antenna selection," *IEEE Trans. Signal Process.*, vol. 53, no. 11, pp. 4091–4103, Nov. 2005.
13. P. Sudarshan, N. B. Mehta, A. F. Molisch, and J. Zhang, "Channel statistics-based RF preprocessing with antenna selection," *IEEE Trans. Wireless Commun.*, vol. 5, no. 12, pp. 3501–3511, Dec. 2006.
14. A. F. Molisch and X. Zhang, "FFT-based hybrid antenna selection schemes for spatially correlated MIMO channels," *IEEE Commun. Lett.*, vol. 8, no. 1, pp. 36–38, Jan. 2004.
15. O. El Ayach, S. Rajagopal, S. Abu-Surra, Z. Pi, and R. W. Heath, "Spatially sparse precoding in millimeter wave MIMO systems," *IEEE Trans. Wireless Commun.*, vol. 13, no. 3, pp. 1499–1513, Mar. 2014.
16. A. Alkhateeb, O. El Ayach, G. Leus, and R. W. Heath, "Channel estimation and hybrid precoding for millimeter wave cellular systems," *IEEE J. Sel. Topics Signal Process.*, vol. 8, no. 5, pp. 831–846, Oct. 2014.
17. X. Yu, J.-C. Shen, J. Zhang, and K. B. Letaief, "Alternating minimization algorithms for hybrid precoding in millimeter wave MIMO systems," *IEEE J. Sel. Topics Signal Process.*, vol. 10, no. 3, pp. 485–500, Apr. 2016.
18. R. Mai, D. H. N. Nguyen, and T. Le-Ngoc, "MMSE hybrid precoder design for millimeter-wave massive MIMO systems," in *Proc. IEEE Wireless Commun. Netw. Conf.*, Doha, Qatar, Apr. 2016.
19. F. Sohrabi and W. Yu, "Hybrid digital and analog beamforming design for large-scale antenna arrays," *IEEE J. Sel. Topics Signal Process.*, vol. 10, no. 3, pp. 501–513, Apr. 2016.
20. A. Alkhateeb and R. W. Heath, "Frequency selective hybrid precoding for limited feedback millimeter wave systems," *IEEE Trans. Commun.*, vol. 64, no. 5, pp. 1801–1818, May 2016.
21. M. H. M. Costa, "Writing on dirty paper," *IEEE Trans. Inf. Theory*, vol. 29, no. 3, pp. 439–441, May 1983.
22. C. B. Peel, "On 'dirty-paper coding'," *IEEE Signal Process. Mag.*, vol. 20, no. 3, pp. 112–113, May 2003.
23. T. Yoo and A. Goldsmith, "On the optimality of multiantenna broadcast scheduling using zero-forcing beamforming," *IEEE J. Sel. Areas Commun.*, vol. 24, no. 3, pp. 528–541, Mar. 2006.
24. J. Q. Wang, D. J. Love, and M. D. Zoltowski, "User selection with zero-forcing beamforming achieves the asymptotically optimal sum rate," *IEEE Trans. Signal Process.*, vol. 56, no. 8, pp. 3713–3726, Aug. 2008.
25. C. B. Peel, B. M. Hochwald, and A. L. Swindlehurst, "A vector-perturbation technique for near-capacity multiantenna multiuser communication - Part I: Channel inversion and regularization," *IEEE Trans. Commun.*, vol. 53, no. 1, pp. 195–202, Jan. 2005.
26. M. Joham, W. Utschick, and J. A. Nossek, "Linear transmit processing in MIMO communications systems," *IEEE Trans. Signal Process.*, vol. 53, no. 8, pp. 2700–2712, Aug. 2005.
27. B. M. Hochwald, C. B. Peel, and A. L. Swindlehurst, "A vector-perturbation technique for near-capacity multiantenna multiuser communication - Part II: Perturbation," *IEEE Trans. Commun.*, vol. 53, no. 3, pp. 537–544, Mar. 2005.
28. D. A. Schmidt, M. Joham, and W. Utschick, "Minimum mean square error vector precoding," *European Trans. Telecommun.*, vol. 19, no. 3, pp. 219–231, Apr. 2008.

29. J. H. Conway and N. J. A. Sloane, *Sphere Packings, Lattices and Groups*, ser. Grundlehren der mathematischen Wissenschaften. New York, NY, USA: Springer-Verlag, 2013.
30. R. R. Muller, D. N. Guo, and A. L. Moustakas, "Vector precoding for wireless MIMO systems and its replica analysis," *IEEE J. Sel. Areas Commun.*, vol. 26, no. 3, pp. 530–540, Apr. 2008.
31. D. J. Ryan, I. B. Collings, I. V. L. Clarkson, and R. W. Heath, "Performance of vector perturbation multiuser MIMO systems with limited feedback," *IEEE Trans. Commun.*, vol. 57, no. 9, pp. 2633–2644, Sep. 2009.
32. J. Choi, "Multiuser precoding with limited cooperation for large-scale MIMO multicell downlink," *IEEE Trans. Wireless Commun.*, vol. 14, no. 3, pp. 1295–1308, Mar. 2015.
33. M. Mazrouei-Sebdani and W. A. Krzymień, "On MMSE vector-perturbation precoding for MIMO broadcast channels with per-antenna-group power constraints," *IEEE Trans. Signal Process.*, vol. 61, no. 15, pp. 3745–3751, Aug. 2013.
34. C. B. Chae, S. H. Kim, and R. W. Heath, "Block diagonalized vector perturbation for multiuser MIMO systems," *IEEE Trans. Wireless Commun.*, vol. 7, no. 11, pp. 4051–4057, Nov. 2008.
35. J. Park, B. Lee, and B. Shim, "A MMSE vector precoding with block diagonalization for multiuser MIMO downlink," *IEEE Trans. Commun.*, vol. 60, no. 2, pp. 569–577, Feb. 2012.
36. F. Liu, L. G. Jiang, and C. He, "Advanced joint transceiver design for block-diagonal geometric-mean-decomposition-based multiuser MIMO systems," *IEEE Trans. Veh. Technol.*, vol. 59, no. 2, pp. 692–703, Feb. 2010.
37. W. Yao, S. Chen, and L. Hanzo, "A transceiver design based on uniform channel decomposition and MBER vector perturbation," *IEEE Trans. Veh. Technol.*, vol. 59, no. 6, pp. 3153–3159, Jul. 2010.
38. C. B. Chae, S. H. Kim, and R. W. Heath, "Network coordinated beamforming for cell-boundary users: Linear and nonlinear approaches," *IEEE J. Sel. Topics Signal Process.*, vol. 3, no. 6, pp. 1094–1105, Dec. 2009.
39. S. P. Boyd and L. Vandenberghe, *Convex Optimization*. Cambridge, UK: Cambridge University Press, 2004.
40. L. Liang, W. Xu, and X. Dong, "Low-complexity hybrid precoding in massive multiuser MIMO systems," vol. 3, no. 6, pp. 653–656, Oct. 2014.
41. W. H. Ni and X. D. Dong, "Hybrid block diagonalization for massive multiuser MIMO systems," *IEEE Trans. Commun.*, vol. 64, no. 1, pp. 201–211, Jan. 2016.
42. J. Choi, D. J. Love, and P. Bidigare, "Downlink training techniques for FDD massive MIMO systems: Open-loop and closed-loop training with memory," *IEEE J. Sel. Topics Signal Process.*, vol. 8, no. 5, pp. 802–814, Oct. 2014.
43. A. Adhikary, J. Nam, J.-Y. Ahn, and G. Caire, "Joint spatial division and multiplexing - The large-scale array regime," *IEEE Trans. Inf. Theory*, vol. 59, no. 10, pp. 6441–6463, Oct. 2013.
44. A. Adhikary, E. Al Safadi, M. K. Samimi, R. Wang, G. Caire, T. S. Rappaport, and A. F. Molisch, "Joint spatial division and multiplexing for mm-wave channels," *IEEE J. Sel. Areas Commun.*, vol. 32, no. 6, pp. 1239–1255, Jun. 2014.
45. A. Liu and V. K. N. Lau, "Hierarchical interference mitigation for massive MIMO cellular networks," *IEEE Trans. Signal Process.*, vol. 62, no. 18, pp. 4786–4797, Sep. 2014.
46. ——, "Phase only RF precoding for massive MIMO systems with limited RF chains," *IEEE Trans. Signal Process.*, vol. 62, no. 17, pp. 4505–4515, Sep. 2014.
47. ——, "Two-stage subspace constrained precoding in massive MIMO cellular systems," *IEEE Trans. Wireless Commun.*, vol. 14, no. 6, pp. 3271–3279, Jun. 2015.
48. H. Shin and J. H. Lee, "Capacity of multiple-antenna fading channels: Spatial fading correlation, double scattering, and keyhole," *IEEE Trans. Inf. Theory*, vol. 49, no. 10, pp. 2636–2647, Oct. 2003.
49. M. Zhang, P. J. Smith, and M. Shafi, "An extended one-ring MIMO channel model," *IEEE Trans. Wireless Commun.*, vol. 6, no. 8, pp. 2759–2764, Aug. 2007.
50. D. Kim, G. Lee, and Y. Sung, "Two-stage beamformer design for massive MIMO downlink by trace quotient formulation," *IEEE Trans. Commun.*, vol. 63, no. 6, pp. 2200–2211, Jun. 2015.
51. A. Liu and V. K. N. Lau, "Two-stage constant-envelope precoding for low-cost massive MIMO systems," *IEEE Trans. Signal Process.*, vol. 64, no. 2, pp. 485–494, Jan. 2016.

52. J. T. Chen and V. K. N. Lau, "Two-tier precoding for FDD multi-cell massive MIMO time-varying interference networks," *IEEE J. Sel. Areas Commun.*, vol. 32, no. 6, pp. 1230–1238, Jun. 2014.
53. Y. Xu, G. Yue, and S. Mao, "User grouping for massive MIMO in FDD systems: New design methods and analysis," *IEEE Access*, vol. 2, pp. 947–959, 2014.
54. J. Nam, A. Adhikary, J.-Y. Ahn, and G. Caire, "Joint spatial division and multiplexing: Opportunistic beamforming, user grouping and simplified downlink scheduling," *IEEE J. Sel. Topics Signal Process.*, vol. 8, no. 5, pp. 876–890, Oct. 2014.
55. J. Nam, Y. J. Ko, and J. Ha, "User grouping of two-stage MU-MIMO precoding for clustered user geometry," *IEEE Commun. Lett.*, vol. 19, no. 8, pp. 1458–1461, Aug. 2015.
56. S. P. Herath, D. H. N. Nguyen, and T. Le-Ngoc, "Vector perturbation precoding for multi-user CoMP downlink transmission," *IEEE Access*, vol. 3, pp. 1491–1502, Sep. 2015.

Chapter 3
Hybrid Precoding and Combining for Massive MIMO Wireless Backhauling

3.1 Introduction

In this chapter, we explore a joint design of two-timescale hybrid precoding and combining for massive multiple-input multiple-output (MIMO) wireless backhaul communications. With the objective of effective capacity maximization, the statistical channel state information (CSI)-based RF and the instantaneous effective CSI-based baseband solutions are jointly derived for the transmitter and receiver. Previous work such as [1, 2], where perfect CSI-based RF and baseband designs were simultaneously obtained as a reconstruction of the optimal counterpart, is no longer applicable. Under the assumption of jointly correlated channels [3–5], the objective function does not have a closed form. Considering traditional MIMO processing at baseband, we address the RF design with nonuniform- and constant-modulus elements.

We establish the optimal structures of RF precoder and combiner where the conclusion in [6] can be viewed as an instance, and reduce the problem of matrix variables to a formulation of discrete eigenmode selection. We propose an intuitive and effective selection criterion by developing an upper bound for the objective function, and develop a semi-definite relaxation (SDR)-based technique to solve the NP-hard combinatorial problem. In the case of constant-modulus RF design, we derive a novel matrix reconstruction formulation based on magnitude least squares (LS), which is further transformed to a problem of approximate joint diagonalization (AJD) of matrices. This makes it possible to exploit simple and low-complexity Jacobi-like algorithms [7, 8] in search of solutions. Compared with [1, 2], the proposed solution technique enjoys low complexity without imposing suboptimal constraint in the RF domain. By separating the RF and baseband designs, our formulation guarantees the optimality of the baseband solutions. We numerically show that the proposed two-timescale hybrid solutions deliver a near-optimal effective rate under various delay-outage constraints while achieving reduced hardware complexity and channel estimation overhead.

T. Le-Ngoc, R. Mai, *Hybrid Massive MIMO Precoding in Cloud-RAN*,
Wireless Networks, https://doi.org/10.1007/978-3-030-02158-0_3

The rest of this chapter is organized as follows. Section 3.2 outlines the system model and formulates the problems for hybrid precoding and combining design for point-to-point massive MIMO wireless backhauling where the statistical queueing constraint is imposed. In Sects. 3.3, and 3.4, low-complexity algorithms are proposed for two-timescale joint hybrid precoding-combining designs with respect to nonuniform- and constant-modulus RF elements. Illustrative results and discussions are presented in Sect. 3.5. Finally, concluding remarks are given in Sect. 3.6.

3.2 System Model and Problem Statement

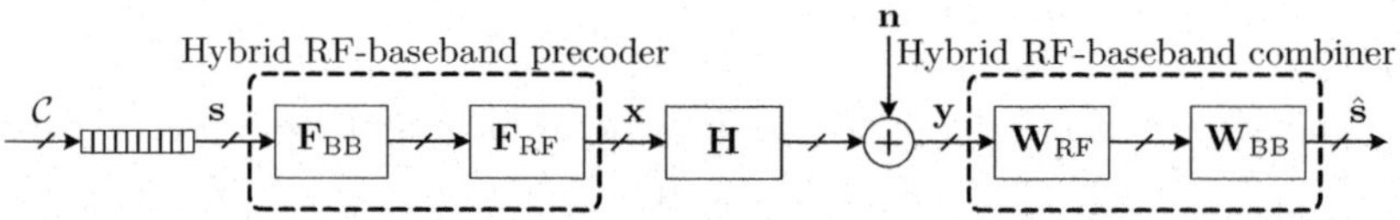

Fig. 3.1 Hybrid RF-baseband precoding and combining subject to statistical queueing constraint

Let us consider a point-to-point wireless backhaul connection over a bandwidth of B, such as one established between a macro-base station (BS) and a micro-BS as defined in Long-Term Evolution (LTE) for relay systems [9]. The transmit BS has $N_\mathrm{t} \gg 1$ antennas associated with $n_\mathrm{t} \ll N_\mathrm{t}$ RF chains and the receive BS has $N_\mathrm{r} \gg 1$ antennas associated with $n_\mathrm{r} \ll N_\mathrm{r}$ RF chains, as illustrated in Fig. 3.1. The receive signal $\mathbf{y}$ is given by

$$\mathbf{y} = \mathbf{H}\mathbf{x} + \mathbf{n},$$

where $\mathbf{H} \in \mathbb{C}^{N_\mathrm{r}\times N_\mathrm{t}}$ is the massive MIMO block-fading channel matrix, $\mathbf{x}$ is the transmit signal, and $\mathbf{n} \sim \mathcal{CN}(\mathbf{0}, \mathbf{I}_{N_\mathrm{r}})$ is the zero mean spatially white Gaussian noise with unit variance. Motivated by the observation that line-of-sight paths are often obstructed due to, e.g., below-the-rooftop micro-BS deployment, we assume non-line-of-sight (NLOS) channels, which can be expressed as [3, 5]

$$\mathbf{H} = \mathbf{U}_\mathrm{r}\tilde{\mathbf{H}}\mathbf{U}_\mathrm{t}^\mathrm{H} = \mathbf{U}_\mathrm{r}\left(\mathbf{M} \odot \mathbf{H}_\mathrm{iid}\right)\mathbf{U}_\mathrm{t}^\mathrm{H} \tag{3.1}$$

where $\mathbf{U}_\mathrm{t} \in \mathbb{C}^{N_\mathrm{t}\times N_\mathrm{t}}$ and $\mathbf{U}_\mathrm{r} \in \mathbb{C}^{N_\mathrm{r}\times N_\mathrm{r}}$ are deterministic unitary matrices, and the pair-wise coupling between each column of $\mathbf{U}_\mathrm{t}$ and $\mathbf{U}_\mathrm{r}$ is characterized by $\tilde{\mathbf{H}} = \mathbf{M} \odot \mathbf{H}_\mathrm{iid}$ with $\mathbf{M}$ deterministic and $\mathbf{H}_\mathrm{iid}$ having independently and identically distributed (i.i.d.) entries (not necessarily Gaussian) with zero mean and unit variance. Further, the variances of the entries of $\tilde{\mathbf{H}}$ can be related to $\mathbf{M}$ as

$$\boldsymbol{\Omega} \triangleq \mathbb{E}\left[\tilde{\mathbf{H}} \odot \tilde{\mathbf{H}}^*\right] = \mathbf{M} \odot \mathbf{M}. \tag{3.2}$$

The matrix $\boldsymbol{\Omega} \in \mathbb{C}^{N_r \times N_t}$ is called the channel coupling matrix (CCM) in that its entries correspond to the coupling power. The channel power gain is normalized as

$$\mathbb{E}\left[\|\mathbf{H}\|_F^2\right] = N_r N_t.$$

Interestingly, the general model in (3.1) can be particularized to some special cases. On one hand, if both link ends are equipped with uniform linear arrays (ULAs), and $\mathbf{U}_t$ and $\mathbf{U}_r$ are constrained to be discrete Fourier transform (DFT) matrices, $\mathbf{H}$ corresponds to the virtual MIMO representation [10]. In this case, the columns of $\mathbf{U}_t$ and $\mathbf{U}_r$ can be interpreted as the physical transmit and receive beamforming directions respectively. On the other hand, if $\mathbf{U}_t$ is chosen as the eigenvectors of the transmit correlation matrix $\mathbf{R}_t \triangleq \mathbb{E}[\mathbf{H}^H\mathbf{H}]$ and $\mathbf{U}_r$ as the eigenvectors of the receive correlation matrix $\mathbf{R}_r \triangleq \mathbb{E}[\mathbf{H}\mathbf{H}^H]$, $\mathbf{H}$ becomes the eigenmode representation [3]. In particular, through the eigenvalue decomposition (EVD), we arrive at

$$\mathbf{R}_t = \mathbf{U}_t \boldsymbol{\Lambda}_t \mathbf{U}_t^H, \tag{3.3}$$

$$\mathbf{R}_r = \mathbf{U}_r \boldsymbol{\Lambda}_r \mathbf{U}_r^H. \tag{3.4}$$

Further, when it is possible to write $[\boldsymbol{\Omega}]_{i,j} = \lambda_{r,i}\lambda_{t,j}$, with $\lambda_{r,i}$ as the ith eigenvalue of $\mathbf{R}_r$ and $\lambda_{t,j}$ as the jth eigenvalue of $\mathbf{R}_t$, the jointly correlated channel model in (3.1) is reduced to a separable correlation channel model, i.e.,

$$\mathbf{H} = \mathbf{U}_r \boldsymbol{\Lambda}_r \mathbf{H}_{\text{iid}} \boldsymbol{\Lambda}_t \mathbf{U}_t^H.$$

For standard Gaussian distributed $\mathbf{H}_{\text{iid}}$, this is the familiar Kronecker channel model [11].

The transmit signal is obtained by linearly precoding N_s data streams $\mathbf{s} \sim \mathcal{CN}(\mathbf{0}, \mathbf{I}_{N_s})$,

$$\mathbf{x} = \mathbf{F}\mathbf{s} = \mathbf{F}_{\text{RF}}\mathbf{F}_{\text{BB}}\mathbf{s}.$$

In particular, the linear precoder $\mathbf{F} = \mathbf{F}_{\text{RF}}\mathbf{F}_{\text{BB}}$ consists of a baseband digital precoder $\mathbf{F}_{\text{BB}} \in \mathbb{C}^{n_t \times N_s}$ followed by an RF analog precoder $\mathbf{F}_{\text{RF}} \in \mathbb{C}^{N_t \times n_t}$ associated with n_t RF chains. In the presence of the RF precoding stage, the dimension of the effective channel $\mathbf{H}\mathbf{F}_{\text{RF}}$ seen by the baseband is substantially reduced, and therefore, it is practically reasonable to assume that the baseband has instantaneous access to the effective channel. On the other hand, on account of signaling delay and channel estimation overhead, it makes sense to assume that only the statistical CSI, which varies on a slow timescale, is available to RF processing. The power constraint is imposed as

$$\mathrm{tr}\left(\mathbb{E}\left[\mathbf{x}\mathbf{x}^{\mathrm{H}}\right]\right) = \|\mathbf{F}_{\mathrm{RF}}\mathbf{F}_{\mathrm{BB}}\|_{\mathrm{F}}^{2} \le P_{\mathrm{t}}.$$

After linear combining at the receiver, the estimated signal is given by

$$\hat{\mathbf{s}} = \mathbf{W}^{\mathrm{H}}\mathbf{y} = \mathbf{W}_{\mathrm{BB}}^{\mathrm{H}}\mathbf{W}_{\mathrm{RF}}^{\mathrm{H}}\mathbf{y},$$

where, similarly to the hybrid precoder, the linear combiner $\mathbf{W} = \mathbf{W}_{\mathrm{RF}}\mathbf{W}_{\mathrm{BB}}$ is comprised by two-timescale, two-stage processing: a statistical CSI-based RF analog combiner $\mathbf{W}_{\mathrm{RF}} \in \mathbb{C}^{N_{\mathrm{r}}\times n_{\mathrm{r}}}$ associated with n_{r} RF chains, and an instantaneous CSI-based baseband digital combiner $\mathbf{W}_{\mathrm{BB}} \in \mathbb{C}^{n_{\mathrm{r}}\times N_{\mathrm{s}}}$. Accordingly, the mean square error (MSE) matrix is defined as

$$\begin{aligned}\mathbf{E} &\triangleq \mathbb{E}\left[\left(\mathbf{s}-\hat{\mathbf{s}}\right)\left(\mathbf{s}-\hat{\mathbf{s}}\right)^{\mathrm{H}}\right] \\ &= \left(\mathbf{I}_{N_{\mathrm{s}}} - \mathbf{W}_{\mathrm{BB}}^{\mathrm{H}}\mathbf{H}_{\mathrm{eff}}\mathbf{F}_{\mathrm{BB}}\right)\left(\mathbf{I}_{N_{\mathrm{s}}} - \mathbf{W}_{\mathrm{BB}}^{\mathrm{H}}\mathbf{H}_{\mathrm{eff}}\mathbf{F}_{\mathrm{BB}}\right)^{\mathrm{H}} + \mathbf{W}_{\mathrm{BB}}^{\mathrm{H}}\mathbf{R}_{n}\mathbf{W}_{\mathrm{BB}},\end{aligned} \tag{3.5}$$

where $\mathbf{H}_{\mathrm{eff}} \triangleq \mathbf{W}_{\mathrm{RF}}^{\mathrm{H}}\mathbf{H}\mathbf{F}_{\mathrm{RF}}$ is the effective channel from the perspective of the baseband, and $\mathbf{R}_{n} \triangleq \mathbf{W}_{\mathrm{RF}}^{\mathrm{H}}\mathbf{W}_{\mathrm{RF}}$ is the noise covariance matrix after RF combining. To minimize the MSE of each data stream (i.e., the diagonal elements of $\mathbf{E}$), it is known that the optimal linear baseband combiner is the linear minimum MSE (MMSE) combiner [12], i.e.,

$$\mathbf{W}_{\mathrm{BB}}^{\mathrm{opt}} \triangleq \left(\mathbf{H}_{\mathrm{eff}}\mathbf{F}_{\mathrm{BB}}\mathbf{F}_{\mathrm{BB}}^{\mathrm{H}}\mathbf{H}_{\mathrm{eff}}^{\mathrm{H}} + \mathbf{I}_{n_{\mathrm{r}}}\right)^{-1}\mathbf{H}_{\mathrm{eff}}\mathbf{F}_{\mathrm{BB}}. \tag{3.6}$$

Substituting (3.6) into (3.5), and using the matrix inversion lemma [13], $\mathbf{E}$ can be simplified as

$$\mathbf{E} = \left(\mathbf{I}_{N_{\mathrm{s}}} + \mathbf{F}_{\mathrm{BB}}^{\mathrm{H}}\mathbf{H}_{\mathrm{eff}}^{\mathrm{H}}\mathbf{R}_{n}^{-1}\mathbf{H}_{\mathrm{eff}}\mathbf{F}_{\mathrm{BB}}\right)^{-1}.$$

Therefore, the instantaneous transmission rate is given by Palomar et al. [12]

$$\begin{aligned}\mathcal{R} &= -B\log_2|\mathbf{E}| \\ &= B\log_2\left|\mathbf{I}_{N_{\mathrm{s}}} + \mathbf{F}_{\mathrm{BB}}^{\mathrm{H}}\mathbf{H}_{\mathrm{eff}}^{\mathrm{H}}\mathbf{R}_{n}^{-1}\mathbf{H}_{\mathrm{eff}}\mathbf{F}_{\mathrm{BB}}\right| \ \mathrm{b/s}.\end{aligned} \tag{3.7}$$

For scenarios such as coordinated beamforming which requires timely CSI exchange or relaying where end-to-end QoS needs to be guaranteed, it is necessary to consider transmission subject to statistical delay constraint. In an effort to characterize the actual throughput, the concept of effective capacity was proposed in [14]. Under the assumption of block fading channels, it is defined as

$$\mathcal{C} = -\frac{1}{\theta T} \ln \mathbb{E}_R \left[e^{-\theta T R} \right] \text{ b/s,} \tag{3.8}$$

where T is the fading block length, θ is the QoS exponent, and R is the rate of a stationary and ergodic stochastic service process. In particular, if we let Q denote the length of a steady-state queue, the QoS exponent θ is related to the buffer-violation (or equivalently, delay-outage) probability by

$$-\theta = \lim_{Q_{\max} \to \infty} \frac{\ln \Pr \{Q > Q_{\max}\}}{Q_{\max}},$$

where $Q_{\max}$ is the maximum allowable queue length. Since $\Pr\{Q > Q_{\max}\} \approx e^{-\theta Q_{\max}}$ for a sufficiently large $Q_{\max}$, it is seen that θ asymptotically represents the exponential decay rate of the buffer-violation probability. For a given $Q_{\max}$, the delay-outage constraint becomes increasingly stringent as θ increases. For a specific application subject to the delay-outage constraint

$$\Pr \{Q > Q_{\max}\} \leq \kappa,$$

the effective capacity describes the maximum constant arrival rate at the transmit buffer that could be supported by the service rate R. In fact, when the queue is in steady state, the average arrival rate is equal to the average service rate, where the effective capacity can also be interpreted as the maximum average throughput subject to the delay-outage constraint [15, 16]. In our case, since the effective CSI $\mathbf{H}_{\text{eff}}$ is assumed to be perfectly known by the baseband, the service rate is given by the instantaneous transmission rate $\mathcal{R}$ in (3.7), and the expectation over the service rate is essentially over the fading channel. Substituting (3.7) into (3.8), we arrive at

$$\begin{aligned} \mathcal{C}(\mathbf{F}_{\text{RF}}, \mathbf{F}_{\text{BB}}, \mathbf{W}_{\text{RF}}) &= -\frac{1}{\theta T} \ln \mathbb{E}_{\mathbf{H}} \left[e^{-\theta T B \log_2 \left| \mathbf{I}_{N_s} + \mathbf{F}_{\text{BB}}^{\text{H}} \mathbf{H}_{\text{eff}}^{\text{H}} \mathbf{R}_n^{-1} \mathbf{H}_{\text{eff}} \mathbf{F}_{\text{BB}} \right|} \right] \\ &= -\frac{1}{\xi} \log_2 \mathbb{E}_{\mathbf{H}} \left[\left| \mathbf{I}_{N_s} + \mathbf{F}_{\text{BB}}^{\text{H}} \mathbf{H}_{\text{eff}}^{\text{H}} \mathbf{R}_n^{-1} \mathbf{H}_{\text{eff}} \mathbf{F}_{\text{BB}} \right|^{-\xi} \right] \text{ b/s/Hz,} \end{aligned}$$

where $\xi = \theta T B \log_2 e$ is the normalized QoS exponent. In comparison with the conventional ergodic capacity, i.e.,

$$\mathcal{C}_{\text{ergodic}} = \mathbb{E}_{\mathbf{H}} \left[\log_2 \left| \mathbf{I}_{N_s} + \mathbf{F}_{\text{BB}}^{\text{H}} \mathbf{H}_{\text{eff}}^{\text{H}} \mathbf{R}_n^{-1} \mathbf{H}_{\text{eff}} \mathbf{F}_{\text{BB}} \right| \right] \text{ b/s/Hz,}$$

which assumes statistical CSI and long-term power allocation at the transmitter, it is not difficult to see that by Jensen's inequality, $\mathcal{C} \leq \mathcal{C}_{\text{ergodic}}$ with equality asymptotically achieved when $\theta \to 0$. In other words, the ergodic capacity represents the maximum transmission rate in the absence of statistical delay constraint. Since the effective throughput under statistical delay constraint is of interest here, we choose the effective capacity as our objective function. Assuming statistical CSI in the RF domain and instantaneous effective CSI at baseband, the optimization problem is thus formulated as

$$\max_{\mathbf{F}_{\mathrm{RF}},\mathbf{W}_{\mathrm{RF}}} -\frac{1}{\xi}\log_2 \mathbb{E}_{\mathbf{H}}\Bigg[\bigg(\max_{\mathbf{F}_{\mathrm{BB}}} \Big|\mathbf{I}_{N_s} + \mathbf{F}_{\mathrm{BB}}^{\mathrm{H}} \underbrace{\mathbf{F}_{\mathrm{RF}}^{\mathrm{H}}\mathbf{H}^{\mathrm{H}}\mathbf{W}_{\mathrm{RF}}}_{=\mathbf{H}_{\mathrm{eff}}^{\mathrm{H}}} \underbrace{\left(\mathbf{W}_{\mathrm{RF}}^{\mathrm{H}}\mathbf{W}_{\mathrm{RF}}\right)^{-1}}_{=\mathbf{R}_n} \times \underbrace{\mathbf{W}_{\mathrm{RF}}^{\mathrm{H}}\mathbf{H}\mathbf{F}_{\mathrm{RF}}}_{=\mathbf{H}_{\mathrm{eff}}} \mathbf{F}_{\mathrm{BB}}\Big|\bigg)^{-\xi}\Bigg]$$

$$\text{s.t.} \quad \|\mathbf{F}_{\mathrm{RF}}\mathbf{F}_{\mathrm{BB}}\|_{\mathrm{F}}^2 \le P_{\mathrm{t}}.$$

The problem statement indicates that we approach hybrid design by first addressing the baseband stage and then the RF stage. Note that the baseband precoder $\mathbf{F}_{\mathrm{BB}}$ is obtained by rate adaptation over each fading block, and expectation over the channel $\mathbf{H}$ suggests that the RF precoder $\mathbf{F}_{\mathrm{RF}}$ and combiner $\mathbf{W}_{\mathrm{RF}}$ lie in the subspace of the channel correlation matrix $\mathbf{R}_{\mathrm{t}}$ and $\mathbf{R}_{\mathrm{r}}$, respectively.

By singular value decomposition (SVD), we express the RF precoder and combiner as $\mathbf{F}_{\mathrm{RF}} = \mathbf{U}_F \boldsymbol{\Sigma}_F \mathbf{V}_F^{\mathrm{H}}$ and $\mathbf{W}_{\mathrm{RF}} = \mathbf{U}_W \boldsymbol{\Sigma}_W \mathbf{V}_W^{\mathrm{H}}$ respectively, where $\mathbf{U}_F \in \mathbb{C}^{N_{\mathrm{t}}\times n_{\mathrm{t}}}$, $\mathbf{U}_W \in \mathbb{C}^{N_{\mathrm{r}}\times n_{\mathrm{r}}}$ are semi-unitary matrices, $\boldsymbol{\Sigma}_F \in \mathbb{C}^{n_{\mathrm{t}}\times n_{\mathrm{t}}}$, $\boldsymbol{\Sigma}_W \in \mathbb{C}^{n_{\mathrm{r}}\times n_{\mathrm{r}}}$ are diagonal matrices with singular values arranged in non-increasing order, and $\mathbf{V}_F \in \mathbb{C}^{n_{\mathrm{t}}\times n_{\mathrm{t}}}$, $\mathbf{V}_W \in \mathbb{C}^{n_{\mathrm{r}}\times n_{\mathrm{r}}}$ are unitary matrices. By substitution, it is straightforward to show that

$$\begin{aligned}\mathbf{H}_{\mathrm{eff}}^{\mathrm{H}}\mathbf{R}_n^{-1}\mathbf{H}_{\mathrm{eff}} &= \mathbf{F}_{\mathrm{RF}}^{\mathrm{H}}\mathbf{H}^{\mathrm{H}}\mathbf{U}_W\mathbf{U}_W^{\mathrm{H}}\mathbf{H}\mathbf{F}_{\mathrm{RF}} \\ &= \mathbf{V}_F\boldsymbol{\Sigma}_F \underbrace{\mathbf{U}_F^{\mathrm{H}}\mathbf{H}^{\mathrm{H}}\mathbf{U}_W\mathbf{U}_W^{\mathrm{H}}\mathbf{H}\mathbf{U}_F}_{\triangleq\mathbf{R}_{\mathrm{eff}}} \boldsymbol{\Sigma}_F\mathbf{V}_F^{\mathrm{H}}.\end{aligned} \tag{3.9}$$

Let $\widetilde{\mathbf{F}}_{\mathrm{BB}} = \boldsymbol{\Sigma}_F\mathbf{V}_F^{\mathrm{H}}\mathbf{F}_{\mathrm{BB}}$. We may reformulate the baseband precoder design as

$$\begin{aligned}\max_{\widetilde{\mathbf{F}}_{\mathrm{BB}}\in\mathbb{C}^{n_{\mathrm{t}}\times N_{\mathrm{s}}}} \quad & \log_2\left|\mathbf{I}_{N_s} + \widetilde{\mathbf{F}}_{\mathrm{BB}}^{\mathrm{H}}\mathbf{R}_{\mathrm{eff}}\widetilde{\mathbf{F}}_{\mathrm{BB}}\right| \\ \text{s.t.} \quad & \|\widetilde{\mathbf{F}}_{\mathrm{BB}}\|_{\mathrm{F}}^2 \le P_{\mathrm{t}}.\end{aligned} \tag{3.10}$$

It was shown in [12] that the optimal solution to (3.10) is given by $\widetilde{\mathbf{F}}_{\mathrm{BB}}^{\mathrm{opt}} = \mathbf{U}_{\mathbf{R}_{\mathrm{eff}}}\boldsymbol{\Sigma}_{\mathrm{BB}}^{\mathrm{opt}}$, where the beamformer $\mathbf{U}_{\mathbf{R}_{\mathrm{eff}}} \in \mathbb{C}^{n_{\mathrm{t}}\times N_{\mathrm{s}}}$ has as columns the eigenvectors of $\mathbf{R}_{\mathrm{eff}}$ associated with the N_{s} largest eigenvalues $\lambda_{\mathbf{R}_{\mathrm{eff}},[1]},\ldots,\lambda_{\mathbf{R}_{\mathrm{eff}},[N_{\mathrm{s}}]}$, and the diagonal matrix $\boldsymbol{\Sigma}_{\mathrm{BB}}^{\mathrm{opt}} = \mathrm{diag}\{\sigma_{\mathrm{BB},1},\ldots,\sigma_{\mathrm{BB},N_{\mathrm{s}}}\}$ performs waterfilling-based power allocation. As a result, $\mathbf{R}_{\mathrm{eff}}$ is diagonalized by $\widetilde{\mathbf{F}}_{\mathrm{BB}}^{\mathrm{opt}}$, and (3.10) is reduced to a scalar optimization problem as

$$\max_{\{\sigma_{\mathrm{BB},i}^2\}} \sum_{i=1}^{N_{\mathrm{s}}} \log_2\left(1+\sigma_{\mathrm{BB},i}^2\lambda_{\mathbf{R}_{\mathrm{eff}},[i]}\right)$$

$$\text{s.t.} \ \sum_{i=1}^{N_s} \sigma_{\mathrm{BB},i}^2 \le P_t.$$

Such concave problem can be easily solved to global optimality by examining the Karush-Kuhn-Tucker (KKT) conditions, which yields the waterfilling solution as

$$\sigma_{\mathrm{BB},i}^2 = \left(\mu^{-1} - \lambda_{\mathbf{R}_{\mathrm{eff}},[i]}^{-1}\right)^+, \ 1 \le i \le N_s, \tag{3.11}$$

with the water-level μ^{-1} satisfying

$$\sum_{i=1}^{N_s} \left(\mu^{-1} - \lambda_{\mathbf{R}_{\mathrm{eff}},[i]}^{-1}\right)^+ = P_t. \tag{3.12}$$

Therefore, we recover the baseband solution from $\widetilde{\mathbf{F}}_{\mathrm{BB}}^{\mathrm{opt}}$ as

$$\mathbf{F}_{\mathrm{BB}}^{\mathrm{opt}} = \mathbf{V}_F \boldsymbol{\Sigma}_F^{-1} \mathbf{U}_{\mathbf{R}_{\mathrm{eff}}} \boldsymbol{\Sigma}_{\mathrm{BB}}^{\mathrm{opt}}. \tag{3.13}$$

We now argue that for any given RF precoder $\mathbf{F}_{\mathrm{RF}}$, the hybrid precoder $\mathbf{F}_{\mathrm{RF}}\mathbf{F}_{\mathrm{BB}}^{\mathrm{opt}}$ is optimal. Suppose that this is not the case, i.e., there exists a baseband precoder $\mathbf{F}_{\mathrm{BB}}^{\mathrm{alt}}$ such that $\mathbf{F}_{\mathrm{RF}}\mathbf{F}_{\mathrm{BB}}^{\mathrm{alt}}$ achieves a higher instantaneous transmission rate. On one hand, following Hadamard inequality, it holds that $\mathbf{F}_{\mathrm{BB}}^{\mathrm{alt}}$ must diagonalize $\mathbf{H}_{\mathrm{eff}}^{\mathrm{H}}\mathbf{R}_n^{-1}\mathbf{H}_{\mathrm{eff}}$, i.e.,

$$(\mathbf{F}_{\mathrm{BB}}^{\mathrm{alt}})^{\mathrm{H}}\mathbf{H}_{\mathrm{eff}}^{\mathrm{H}}\mathbf{R}_n^{-1}\mathbf{H}_{\mathrm{eff}}\mathbf{F}_{\mathrm{BB}}^{\mathrm{alt}} = \mathbf{D}_{\mathrm{alt}},$$

where $\mathbf{D}_{\mathrm{alt}} \in \mathbb{C}^{N_s \times N_s}$ is diagonal. Using (3.9), we also have

$$(\mathbf{F}_{\mathrm{BB}}^{\mathrm{alt}})^{\mathrm{H}}\mathbf{F}_{\mathrm{RF}}^{\mathrm{H}}\mathbf{H}^{\mathrm{H}}\mathbf{U}_W\mathbf{U}_W^{\mathrm{H}}\mathbf{H}\mathbf{F}_{\mathrm{RF}}\mathbf{F}_{\mathrm{BB}}^{\mathrm{alt}} = (\mathbf{F}_{\mathrm{BB}}^{\mathrm{alt}})^{\mathrm{H}}\mathbf{V}_F\boldsymbol{\Sigma}_F\mathbf{R}_{\mathrm{eff}}\boldsymbol{\Sigma}_F\mathbf{V}_F^{\mathrm{H}}\mathbf{F}_{\mathrm{BB}}^{\mathrm{alt}} = \mathbf{D}_{\mathrm{alt}}.$$

Based on [12, Lemma 12], there exists $\mathbf{F}_{\mathrm{BB}}^{\mathrm{eig}} = \mathbf{U}_{\mathbf{R}_{\mathrm{eff}}}\boldsymbol{\Sigma}_{\mathrm{BB}}^{\mathrm{eig}}$ with $\boldsymbol{\Sigma}_{\mathrm{BB}}^{\mathrm{eig}} \in \mathbb{C}^{N_s \times N_s}$ diagonal such that

$$(\mathbf{F}_{\mathrm{BB}}^{\mathrm{eig}})^{\mathrm{H}}\mathbf{R}_{\mathrm{eff}}\mathbf{F}_{\mathrm{BB}}^{\mathrm{eig}} = \boldsymbol{\Sigma}_{\mathrm{BB}}^{\mathrm{eig}}\mathbf{U}_{\mathbf{R}_{\mathrm{eff}}}^{\mathrm{H}}\mathbf{R}_{\mathrm{eff}}\mathbf{U}_{\mathbf{R}_{\mathrm{eff}}}\boldsymbol{\Sigma}_{\mathrm{BB}}^{\mathrm{eig}} = \mathbf{D}_{\mathrm{alt}} \tag{3.14}$$

while $\|\mathbf{F}_{\mathrm{BB}}^{\mathrm{eig}}\|_{\mathrm{F}}^2 = \mathrm{tr}[(\boldsymbol{\Sigma}_{\mathrm{BB}}^{\mathrm{eig}})^2] \le \|\boldsymbol{\Sigma}_F\mathbf{V}_F^{\mathrm{H}}\mathbf{F}_{\mathrm{BB}}^{\mathrm{alt}}\|_{\mathrm{F}}^2 = \|\mathbf{F}_{\mathrm{RF}}\mathbf{F}_{\mathrm{BB}}^{\mathrm{alt}}\|_{\mathrm{F}}^2$. In other words, $\mathbf{F}_{\mathrm{BB}}^{\mathrm{eig}}$ achieves a better rate performance than $\mathbf{F}_{\mathrm{RF}}\mathbf{F}_{\mathrm{BB}}^{\mathrm{alt}}$. On the other hand, from (3.13) we have

$$(\mathbf{F}_{\mathrm{BB}}^{\mathrm{opt}})^{\mathrm{H}}\mathbf{F}_{\mathrm{RF}}^{\mathrm{H}}\mathbf{H}^{\mathrm{H}}\mathbf{U}_W\mathbf{U}_W^{\mathrm{H}}\mathbf{H}\mathbf{F}_{\mathrm{RF}}\mathbf{F}_{\mathrm{BB}}^{\mathrm{opt}} = \boldsymbol{\Sigma}_{\mathrm{BB}}^{\mathrm{opt}}\mathbf{U}_{\mathbf{R}_{\mathrm{eff}}}^{\mathrm{H}}\mathbf{R}_{\mathrm{eff}}\mathbf{U}_{\mathbf{R}_{\mathrm{eff}}}\boldsymbol{\Sigma}_{\mathrm{BB}}^{\mathrm{opt}} \tag{3.15}$$

and $\|\mathbf{F}_{\text{RF}}\mathbf{F}_{\text{BB}}^{\text{opt}}\|_{\text{F}}^{2} = \text{tr}[(\boldsymbol{\Sigma}_{\text{BB}}^{\text{opt}})^{2}]$. Comparing (3.14) with (3.15), we see that subject to the same power constraint $\text{tr}[(\boldsymbol{\Sigma}_{\text{BB}}^{\text{opt}})^{2}] = \text{tr}[(\boldsymbol{\Sigma}_{\text{BB}}^{\text{eig}})^{2}] = P_{\text{t}}$, the waterfilling-based $\boldsymbol{\Sigma}_{\text{BB}}^{\text{opt}}$ necessarily delivers a better performance than $\boldsymbol{\Sigma}_{\text{BB}}^{\text{eig}}$, i.e., $\mathbf{F}_{\text{RF}}\mathbf{F}_{\text{BB}}^{\text{opt}}$ outperforms $\mathbf{F}_{\text{BB}}^{\text{eig}}$ and thus $\mathbf{F}_{\text{RF}}\mathbf{F}_{\text{BB}}^{\text{alt}}$. This contradicts the assumption of the superiority of $\mathbf{F}_{\text{RF}}\mathbf{F}_{\text{BB}}^{\text{alt}}$. In light of (3.13), the effective capacity is thus reduced to

$$\mathcal{C}\left(\mathbf{F}_{\text{RF}}, \mathbf{W}_{\text{RF}}\right) = -\frac{1}{\xi}\log_2 \mathbb{E}_{\mathbf{H}}\left[\prod_{i=1}^{N_{\text{s}}}\left(1+\sigma_i^2\lambda_{\mathbf{R}_{\text{eff}},[i]}\right)^{-\xi}\right].$$

It is seen that the effective capacity depends on the eigenvalues of $\mathbf{R}_{\text{eff}} = \mathbf{U}_F^{\text{H}}\mathbf{H}^{\text{H}}\mathbf{U}_W\mathbf{U}_W^{\text{H}}\mathbf{H}\mathbf{U}_F$, which is in turn dependent on the left singular vectors $\mathbf{U}_F$ of the RF precoder $\mathbf{F}_{\text{RF}}$ and $\mathbf{U}_W$ of the RF combiner $\mathbf{W}_{\text{RF}}$. The implication is that it suffices to consider semi-unitary RF precoders and combiners, which are posed as solutions to the following stochastic optimization problem

$$\begin{aligned}\max_{\mathbf{F}_{\text{RF}},\mathbf{W}_{\text{RF}}} \quad & -\frac{1}{\xi}\log_2 \mathbb{E}_{\mathbf{H}}\left[\prod_{i=1}^{N_{\text{s}}}\left(1+\sigma_i^2\lambda_{\mathbf{R}_{\text{eff}},[i]}\right)^{-\xi}\right] \\ \text{s.t.} \quad & \mathbf{F}_{\text{RF}}^{\text{H}}\mathbf{F}_{\text{RF}} = \mathbf{I}_{n_{\text{t}}},\ \mathbf{W}_{\text{RF}}^{\text{H}}\mathbf{W}_{\text{RF}} = \mathbf{I}_{n_{\text{r}}},\end{aligned} \tag{3.16}$$

where we accordingly rewrite $\mathbf{R}_{\text{eff}} = \mathbf{F}_{\text{RF}}^{\text{H}}\mathbf{H}^{\text{H}}\mathbf{W}_{\text{RF}}\mathbf{W}_{\text{RF}}^{\text{H}}\mathbf{H}\mathbf{F}_{\text{RF}}$. We observe that multiplying $\mathbf{F}_{\text{RF}}$ or $\mathbf{W}_{\text{RF}}$ with arbitrary unitary matrices does not affect either the eigenvalues of $\mathbf{R}_{\text{eff}}$ or the semi-unitary constraints. In other words, the RF precoding and combining solutions to (3.16) are in fact subspaces spanned by the columns of $\mathbf{F}_{\text{RF}}$ and $\mathbf{W}_{\text{RF}}$ respectively.

Note that the elements of the representative $\mathbf{F}_{\text{RF}}$ and $\mathbf{W}_{\text{RF}}$ are generally complex numbers of arbitrary magnitudes, which are often implemented using power amplifiers and phase-shifters. For practicality, it is desirable to employ only phase-shifters in the RF domain. Mathematically, this translates to RF elements of uniform magnitudes. By imposing this constraint on (3.16), we formulate the problem of constant-modulus RF design as

$$\begin{aligned}\max_{\mathbf{F}_{\text{RF}},\mathbf{W}_{\text{RF}}} \quad & -\frac{1}{\xi}\log_2 \mathbb{E}_{\mathbf{H}}\left[\prod_{i=1}^{N_{\text{s}}}\left(1+\sigma_i^2\lambda_{\mathbf{R}_{\text{eff}},[i]}\right)^{-\xi}\right] \\ \text{s.t.} \quad & \mathbf{F}_{\text{RF}}^{\text{H}}\mathbf{F}_{\text{RF}} = \mathbf{I}_{n_{\text{t}}},\ \mathbf{F}_{\text{RF}} \in \mathcal{F}_{\text{RF}} \\ & \mathbf{W}_{\text{RF}}^{\text{H}}\mathbf{W}_{\text{RF}} = \mathbf{I}_{n_{\text{r}}},\ \mathbf{W}_{\text{RF}} \in \mathcal{W}_{\text{RF}},\end{aligned} \tag{3.17}$$

where the set of constant-modulus RF precoders is defined as $\mathcal{F}_{\text{RF}} \triangleq \{\mathbf{F}_{\text{RF}} \in \mathbb{C}^{N_{\text{t}}\times n_{\text{t}}} | [\mathbf{F}_{\text{RF}}]_{i,j} = \frac{1}{\sqrt{N_{\text{t}}}}e^{\jmath\phi_{i,j}}\}$, and $\mathcal{W}_{\text{RF}} \triangleq \{\mathbf{W}_{\text{RF}} \in \mathbb{C}^{N_{\text{r}}\times n_{\text{r}}} | [\mathbf{W}_{\text{RF}}]_{i,j} = \frac{1}{\sqrt{N_{\text{r}}}}e^{\jmath\psi_{i,j}}\}$ for the set of constant-modulus RF combiners. One major challenge of optimizing (3.16) and (3.17) is that a closed-form objective is not available in the absence of channel distribution. In the case of constant-modulus design (3.17), the non-convex modulus constraints make it even more difficult to directly find an optimal solution. In Sect. 3.3, we study the optimal structures of $\mathbf{F}_{\mathrm{RF}}$ and $\mathbf{W}_{\mathrm{RF}}$ in (3.16) and propose an approximate reformulation which can be efficiently solved by SDR. Exploiting the observation that the precoding and combining solutions are actually subspaces, we propose a low-complexity matrix reconstruction algorithm to solve the constant-modulus design (3.17) in Sect. 3.4.

3.3 Joint Nonuniform-Modulus RF Tx-Rx Design

In this section, we show that it is possible to derive the optimal structures of $\mathbf{F}_{\mathrm{RF}}$ and $\mathbf{W}_{\mathrm{RF}}$ without knowing a closed-form expression for the objective function. This insight leads to an eigenmode selection problem with respect to the CCM $\boldsymbol{\Omega}$, and a selection criterion is then developed by establishing an upper bound for the objective function. We formulate it as a Boolean quadratic programming problem with linear constraints, and obtain an approximately optimal solution by an SDR technique. To begin with, the following theorem establishes the optimal structures of RF precoding and combining.

Theorem 3.1 *To maximize the effective capacity C, the optimal RF precoder can be expressed as $\mathbf{F}_{\mathrm{RF}}^{\mathrm{opt}} = \mathbf{U}_{\mathrm{t}}\mathbf{S}_F$, where $\mathbf{U}_{\mathrm{t}} \in \mathbb{C}^{N_{\mathrm{t}} \times N_{\mathrm{t}}}$ is the transmit eigenmodes in (3.3), and $\mathbf{S}_F \in \mathbb{C}^{N_{\mathrm{t}} \times n_{\mathrm{t}}}$ is a selection matrix that picks n_{t} columns out of $\mathbf{U}_{\mathrm{t}}$. Similarly, the optimal RF combiner is $\mathbf{W}_{\mathrm{RF}}^{\mathrm{opt}} = \mathbf{U}_{\mathrm{r}}\mathbf{S}_W$, where $\mathbf{U}_{\mathrm{r}} \in \mathbb{C}^{N_{\mathrm{r}} \times N_{\mathrm{r}}}$ is the receive eigenmodes in (3.4), and $\mathbf{S}_W \in \mathbb{C}^{N_{\mathrm{r}} \times n_{\mathrm{r}}}$ is a selection matrix.*

Proof Here we outline the main idea of the proof and leave the mathematical details to Appendix 1. The feasibility of deriving the optimal structures of $\mathbf{F}_{\mathrm{RF}}$ and $\mathbf{W}_{\mathrm{RF}}$ without a closed-form objective hinges on three observations: (1) the objective function C is concave in $\mathbf{U}_{\mathrm{t}}^{\mathrm{H}}\mathbf{F}_{\mathrm{RF}}\mathbf{F}_{\mathrm{RF}}^{\mathrm{H}}\mathbf{U}_{\mathrm{t}}$ and $\mathbf{U}_{\mathrm{r}}^{\mathrm{H}}\mathbf{W}_{\mathrm{RF}}\mathbf{W}_{\mathrm{RF}}^{\mathrm{H}}\mathbf{U}_{\mathrm{r}}$; (2) the elements of $\mathbf{H}_{\mathrm{iid}}$ are independent and symmetrically distributed around the origin by assumption, and therefore changing the signs of the elements does not affect their distribution; (3) it suffices to consider semi-unitary RF precoder and combiner, i.e., $\mathbf{F}_{\mathrm{RF}}^{\mathrm{H}}\mathbf{F}_{\mathrm{RF}} = \mathbf{I}_{n_{\mathrm{t}}}$ and $\mathbf{W}_{\mathrm{RF}}^{\mathrm{H}}\mathbf{W}_{\mathrm{RF}} = \mathbf{I}_{n_{\mathrm{r}}}$. The first and second observations lead to the conclusion that the objective function C is maximized when $\mathbf{U}_{\mathrm{t}}^{\mathrm{H}}\mathbf{F}_{\mathrm{RF}}\mathbf{F}_{\mathrm{RF}}^{\mathrm{H}}\mathbf{U}_{\mathrm{t}}$ and $\mathbf{U}_{\mathrm{r}}^{\mathrm{H}}\mathbf{W}_{\mathrm{RF}}\mathbf{W}_{\mathrm{RF}}^{\mathrm{H}}\mathbf{U}_{\mathrm{r}}$ are diagonal. Based on the third observation and the fact that $\mathbf{U}_{\mathrm{t}}$ and $\mathbf{U}_{\mathrm{r}}$ are semi-unitary, we further conclude that the RF precoder and combiner must be of the form $\mathbf{F}_{\mathrm{RF}} = \mathbf{U}_{\mathrm{t}}\mathbf{S}_F$, $\mathbf{W}_{\mathrm{RF}} = \mathbf{U}_{\mathrm{r}}\mathbf{S}_W$ where $\mathbf{S}_F$ and $\mathbf{S}_W$ are selection matrices.

Remark 3.1 As explained in Sect. 3.2, the entries of the coupling channel in the DFT beamforming space are independently and symmetrically distributed around zero when both transmitter and receiver are equipped with ULA. Thus, the result given by Theorem 3.1 is also applicable to the virtual MIMO representation [10].

Due to coupling between the transmit and receive eigenmodes, it is necessary to jointly optimize $\mathbf{S}_F$ and $\mathbf{S}_W$ for optimal performance. Not surprisingly, in the case of separable correlation, the optimal solutions can be derived separately.

Corollary 3.1 *In the case of separable correlation, the effective capacity* $\mathcal{C}$ *is maximized when the selection matrix* $\mathbf{S}_F$ *for RF precoding picks the transmit eigenmodes corresponding to the* n_t *largest eigenvalues of* $\mathbf{R}_t$ *and the selection matrix* $\mathbf{S}_W$ *for RF combining picks the receive eigenmodes corresponding to the* n_r *largest eigenvalues of* $\mathbf{R}_r$.

Proof See Appendix 2.

Using the conclusion from Theorem 3.1 and the channel model (3.1), we simplify $\mathbf{R}_{\text{eff}}$ as

$$\begin{aligned}\mathbf{R}_{\text{eff}} &= \mathbf{F}_{\text{RF}}^{\text{H}}\mathbf{H}^{\text{H}}\mathbf{W}_{\text{RF}}\mathbf{W}_{\text{RF}}^{\text{H}}\mathbf{H}\mathbf{F}_{\text{RF}} \\ &= \mathbf{S}_F^{\text{H}}\tilde{\mathbf{H}}^{\text{H}}\mathbf{S}_W\mathbf{S}_W^{\text{H}}\tilde{\mathbf{H}}\mathbf{S}_F.\end{aligned}$$

Now the problem boils down to choosing n_r rows $\{i_1, \ldots, i_{n_r}\}$ and n_t columns $\{j_1, \ldots, j_{n_t}\}$ out of the coupling matrix $\tilde{\mathbf{H}}$ such that the effective capacity is maximized. Due to the lack of knowledge on the distribution of $\tilde{\mathbf{H}}$, we approach this problem by deriving an upper bound to $\mathcal{C}$ which reveals an intuitive selection criterion.

Theorem 3.2 *The effective capacity* $\mathcal{C}$ *can be upper bounded by*

$$\mathcal{C}\left(\mathbf{S}_F, \mathbf{S}_W\right) \le \mathcal{C}_{\text{ub}}\left(\mathbf{S}_F, \mathbf{S}_W\right) = N_s \log_2 \frac{P_t}{N_s} - n_t \log_2 n_t + n_t \log_2 \sum_{l=1}^{n_r}\sum_{k=1}^{n_t} [\boldsymbol{\Omega}]_{i_l, j_k},$$

where $\boldsymbol{\Omega}$ *is the CCM in (3.2). Therefore, the Tx-Rx eigenmode pairs are chosen in such a way that the sum of the corresponding channel coupling power is maximized.*

Proof See Appendix 3.

Remark 3.2 For separable correlation, the optimal transmit and receive RF beamformers can be derived without considering the normalized QoS exponent ξ as established in Corollary 3.1. On the contrary, in the case of jointly correlated channels, the selection matrices $\mathbf{S}_F$ and $\mathbf{S}_W$ can depend on ξ. However, because $\mathcal{C}$ cannot be evaluated to a closed form, such design becomes mathematically intractable. Instead, we develop an upper bound $\mathcal{C}_{\text{ub}}$ as an approximation for $\mathcal{C}$ in the high SNR regime irrespective of ξ. Although such upper bound only leads to suboptimal $\mathbf{S}_F$ and $\mathbf{S}_W$, simulation results show that the hybrid solutions with only a limited number of RF chains are robust to the effect of delay-outage constraints when directional scattering is observed.

For an intuitive justification, let us consider the separable correlation channel model. Recall that in this case, $[\boldsymbol{\Omega}]_{i_l, j_k} = \lambda_{r,i_l}\lambda_{t,j_k}$ and accordingly, the upper bound is reduced to

$$\mathcal{C}_{\text{ub}}\left(\mathbf{S}_F, \mathbf{S}_W\right) = N_s \log_2 \frac{P_t}{N_s} - n_t \log_2 n_t + n_t \log_2 \sum_{l=1}^{n_r} \lambda_{r,i_l} \sum_{k=1}^{n_t} \lambda_{t,j_k}.$$

Obviously, this upper bound is maximized when the maximum n_r eigenvalues of $\mathbf{R}_r$ and the maximum n_t eigenvalues of $\mathbf{R}_t$ are selected. This is indeed consistent with the conclusion from Corollary 3.1. Consequently, the problem of (3.16) is recast as

$$\begin{aligned}
&\max_{\{x_i, y_j\}} \sum_{i=1}^{N_r}\sum_{j=1}^{N_t} [\boldsymbol{\Omega}]_{i,j}\, x_i y_j \\
&\text{s.t.} \sum_{i=1}^{N_r} x_i = n_r, \ \sum_{j=1}^{N_t} y_j = n_t \\
&\qquad x_1, \ldots, x_{N_r}, y_1, \ldots, y_{N_t} \in \{0, 1\}.
\end{aligned}$$

The indefinite quadratic objective function renders the problem non-convex, which is actually NP-hard as a result of the Boolean constraints. In light of the large dimensions of $\boldsymbol{\Omega}$, a brute-force search would be computationally infeasible. Instead, we exploit SDR for problem convexification.

Let us redefine the optimization variable as $\mathbf{z} = [z_1, \ldots, z_{N_r+N_t}]^{\mathrm{T}}$ where $z_k = x_k$ for $1 \le k \le N_r$ and $z_k = y_{k-N_r}$ for $N_r + 1 \le k \le N_r + N_t$. Accordingly, the linear constraints become

$$\sum_{k=1}^{N_r} z_k - n_r = 0, \ \sum_{k=N_r+1}^{N_r+N_t} z_k - n_t = 0. \tag{3.18}$$

For the purpose of SDR, define a rank-1 positive semi-definite (PSD) matrix $\mathbf{Z} = \mathbf{z}\mathbf{z}^{\mathrm{T}}$ with the (i, j)th entry as $Z_{ij} = z_i z_j$. Further, multiply the first constraint in (3.18) by z_i, $1 \le i \le N_r$, and the second by z_i, $N_r + 1 \le i \le N_r + N_t$. By replacing $z_i z_k$ by Z_{ik} and z_i by Z_{ii}, we arrive at

$$\sum_{k=1}^{N_r} Z_{ik} - n_r Z_{ii} = 0, \ 1 \le i \le N_r, \ \sum_{k=N_r+1}^{N_r+N_t} Z_{ik} - n_t Z_{ii} = 0, \ N_r + 1 \le i \le N_r + N_t.$$

Relaxing the binary constraint $\mathbf{z} \in \{0, 1\}^{N_r+N_t}$ as $\mathbf{0} \preceq \mathbf{z} \preceq \mathbf{1}$ gives $\mathbf{Z} - \operatorname{diag}(\mathbf{Z}) \operatorname{diag}(\mathbf{Z})^{\mathrm{T}} \succeq 0$. Using Schur complement [17], this is equivalent to $\begin{bmatrix} 1 & \operatorname{diag}(\mathbf{Z})^{\mathrm{T}} \\ \operatorname{diag}(\mathbf{Z}) & \mathbf{Z} \end{bmatrix} \succeq 0$. Finally, dropping the rank-1 constraint on $\mathbf{Z}$, we convexify the original problem as

$$\begin{aligned}
&\max_{\mathbf{Z}} \frac{1}{2}\operatorname{tr}(\mathbf{P}\mathbf{Z}) \\
&\text{s.t.} \sum_{k=1}^{N_r} Z_{ik} - n_r Z_{ii} = 0, \ 1 \le i \le N_r, \ \sum_{k=N_r+1}^{N_r+N_t} Z_{jk} - n_t Z_{jj} = 0, \ 1 \le j \le N_t \\
&\qquad \begin{bmatrix} 1 & \operatorname{diag}(\mathbf{Z})^{\mathrm{T}} \\ \operatorname{diag}(\mathbf{Z}) & \mathbf{Z} \end{bmatrix} \succeq 0,
\end{aligned} \tag{3.19}$$

where $\mathbf{P} = \begin{bmatrix} \mathbf{O} & \boldsymbol{\Omega} \\ \boldsymbol{\Omega}^{\mathrm{T}} & \mathbf{O} \end{bmatrix}$ is symmetric. Convex problems as such can be readily solved by off-the-shelf optimization software packages, e.g., cvx [18]. Once the optimal $\mathbf{Z}^{\mathrm{opt}}$ to (3.19) is acquired, we rearrange Z_{ii}, $1 \le i \le N_{\mathrm{r}}$ in non-increasing order, i.e., $Z_{i_1 i_1} \ge \cdots \ge Z_{i_{N_{\mathrm{r}}} i_{N_{\mathrm{r}}}}$, and similarly for Z_{jj}, $N_{\mathrm{r}} + 1 \le j \le N_{\mathrm{r}} + N_{\mathrm{t}}$ as $Z_{j_1 j_1} \ge \cdots \ge Z_{j_{N_{\mathrm{t}}} j_{N_{\mathrm{t}}}}$. Subsequently, a feasible $\mathbf{x}^{\mathrm{opt}} \triangleq [x_1^{\mathrm{opt}}, \ldots, x_{N_{\mathrm{r}}}^{\mathrm{opt}}]^{\mathrm{T}}$ and $\mathbf{y}^{\mathrm{opt}} \triangleq [y_1^{\mathrm{opt}}, \ldots, y_{N_{\mathrm{t}}}^{\mathrm{opt}}]^{\mathrm{T}}$ are heuristically obtained as [19]

$$x_k^{\mathrm{opt}} = \begin{cases} 1, & k = i_1, \ldots, i_{n_{\mathrm{r}}} \\ 0, & \text{otherwise} \end{cases}, \quad y_k^{\mathrm{opt}} = \begin{cases} 1, & k = j_1 - N_{\mathrm{r}}, \ldots, j_{n_{\mathrm{t}}} - N_{\mathrm{r}} \\ 0, & \text{otherwise} \end{cases}. \tag{3.20}$$

3.4 RF Phase-Shifting Design by Matrix Reconstruction

The lack of a closed-form objective and the non-convexity of the feasible sets $\mathcal{F}_{\mathrm{RF}}$ and $\mathcal{W}_{\mathrm{RF}}$ make it difficult to solve (3.17) to optimality. Because a direct approach to (3.17) seems elusive, here we propose a suboptimal matrix reconstruction approach. In Sect. 3.2, it is observed that the RF analog designs are unitary-invariant. In other words, for optimal $\mathbf{F}_{\mathrm{RF}}^{\mathrm{opt}}$ and $\mathbf{W}_{\mathrm{RF}}^{\mathrm{opt}}$, $\mathbf{F}_{\mathrm{RF}}^{\mathrm{opt}}\mathbf{B}_F$ and $\mathbf{W}_{\mathrm{RF}}^{\mathrm{opt}}\mathbf{B}_W$ are also optimal for arbitrary unitary matrices $\mathbf{B}_F$ and $\mathbf{B}_W$. The idea is to exploit $\mathbf{B}_F$ and $\mathbf{B}_W$ so that the resulting $\mathbf{F}_{\mathrm{RF}}^{\mathrm{cm}} = \mathbf{F}_{\mathrm{RF}}^{\mathrm{opt}}\mathbf{B}_F$ and $\mathbf{W}_{\mathrm{RF}}^{\mathrm{cm}} = \mathbf{W}_{\mathrm{RF}}^{\mathrm{opt}}\mathbf{B}_W$ satisfy the constant-modulus constraints. To this end, a magnitude LS formulation is developed in Sect. 3.4.1, which is shown to be solvable by low-complexity AJD algorithms in Sect. 3.4.2.

3.4.1 Magnitude LS Formulation

For constant-modulus RF precoding design, the proposed matrix reconstruction approach is built upon solving the following optimization problem

$$\begin{aligned} \min_{\mathbf{F}_{\mathrm{RF}}^{\mathrm{cm}}, \mathbf{B}_F} \quad & \left\| \mathbf{F}_{\mathrm{RF}}^{\mathrm{opt}}\mathbf{B}_F - \mathbf{F}_{\mathrm{RF}}^{\mathrm{cm}} \right\|_{\mathrm{F}}^2 \\ \text{s.t.} \quad & \mathbf{F}_{\mathrm{RF}}^{\mathrm{cm}} \in \mathcal{F}_{\mathrm{RF}}, \mathbf{B}_F^{\mathrm{H}}\mathbf{B}_F = \mathbf{I}_{n_{\mathrm{t}}}, \end{aligned} \tag{3.21}$$

and analogously, for constant-modulus RF combining design

$$\begin{aligned} \min_{\mathbf{W}_{\mathrm{RF}}^{\mathrm{cm}}, \mathbf{B}_W} \quad & \left\| \mathbf{F}_{\mathrm{RF}}^{\mathrm{opt}}\mathbf{B}_W - \mathbf{W}_{\mathrm{RF}}^{\mathrm{cm}} \right\|_{\mathrm{F}}^2 \\ \text{s.t.} \quad & \mathbf{W}_{\mathrm{RF}}^{\mathrm{cm}} \in \mathcal{W}_{\mathrm{RF}}, \mathbf{B}_W^{\mathrm{H}}\mathbf{B}_W = \mathbf{I}_{n_{\mathrm{r}}}. \end{aligned} \tag{3.22}$$

Obviously, exact reconstruction is achievable when the optimal values of (3.21) and (3.22) are zero. In that the two problems are similar to each other, in what follows, we focus our presentation on RF precoding design.

Although the original objective function in (3.17) is now replaced by the simpler Euclidean distance, the problem is still not easy to solve when constrained to the non-convex constant-modulus set $\mathcal{F}_{\mathrm{RF}}$. In an attempt to subsume the constraint into the objective function, we expand it as

$$\mathcal{J}_{\mathrm{Tx}}\left(\mathbf{F}_{\mathrm{RF}}^{\mathrm{cm}}, \mathbf{B}_F\right) \triangleq \left\|\mathbf{F}_{\mathrm{RF}}^{\mathrm{opt}}\mathbf{B}_F - \mathbf{F}_{\mathrm{RF}}^{\mathrm{cm}}\right\|_{\mathrm{F}}^2 = \sum_{k=1}^{n_{\mathrm{t}}} \left\|\mathbf{F}_{\mathrm{RF}}^{\mathrm{opt}}\mathbf{b}_k - \mathbf{f}_k\right\|^2,$$

where $\mathbf{b}_k$ and $\mathbf{f}_k^{\mathrm{cm}}$ are the ith columns of $\mathbf{B}_F$ and $\mathbf{F}_{\mathrm{RF}}^{\mathrm{cm}}$ respectively. Let $\{\mathbf{f}_{\mathrm{opt},m}^{\mathrm{H}}\}$ be the rows of $\mathbf{F}_{\mathrm{RF}}^{\mathrm{opt}}$, and define

$$\mathcal{L}_k\left(\mathbf{b}_k, \mathbf{f}_k^{\mathrm{cm}}\right) \triangleq \left\|\mathbf{F}_{\mathrm{RF}}^{\mathrm{opt}}\mathbf{b}_k - \mathbf{f}_k^{\mathrm{cm}}\right\|^2 = \sum_{m=1}^{N_{\mathrm{t}}} \left|\mathbf{f}_{\mathrm{opt},m}^{\mathrm{H}}\mathbf{b}_k - e^{j\phi_{m,k}}\right|^2.$$

To minimize $\mathcal{L}_k(\mathbf{b}_k, \mathbf{f}_k^{\mathrm{cm}})$, it can be seen that $\phi_{m,k} = \angle(\mathbf{f}_{\mathrm{opt},m}^{\mathrm{H}}\mathbf{b}_k)$ is the optimal choice. Therefore, we have

$$\mathcal{L}_k\left(\mathbf{b}_k\right) = \sum_{m=1}^{N_{\mathrm{t}}} \left(\left|\mathbf{f}_{\mathrm{opt},m}^{\mathrm{H}}\mathbf{b}_k\right| - 1\right)^2. \tag{3.23}$$

However, this problem is still non-convex due to the magnitude inside the square. Using the bound $(x-1)^2 \le \left(x^2-1\right)^2$, $x \ge 0$, (3.23) can be upper-bounded as

$$\mathcal{L}_k\left(\mathbf{b}_k\right) \le \mathcal{L}_k^{\mathrm{ub}}\left(\mathbf{b}_k\right) \triangleq \sum_{m=1}^{N_{\mathrm{t}}} \left(\left|\mathbf{f}_{\mathrm{opt},m}^{\mathrm{H}}\mathbf{b}_k\right|^2 - 1\right)^2. \tag{3.24}$$

This bound is tight when $\left|\mathbf{f}_{\mathrm{opt},m}^{\mathrm{H}}\mathbf{b}_k\right|$ is in the neighborhood of unity. By linearizing the quadratic term $\left|\mathbf{f}_{\mathrm{opt},m}^{\mathrm{H}}\mathbf{b}_k\right|^2$ in (3.24), we ultimately arrive at a formulation of joint matrix diagonalization.

Since $\mathbf{b}_k^{\mathrm{H}}\mathbf{b}_k = 1$ by assumption, it follows that

$$\begin{aligned}\mathcal{L}_k^{\mathrm{ub}}\left(\mathbf{b}_k\right) &= \sum_{m=1}^{N_{\mathrm{t}}} \left(\left|\mathbf{f}_{\mathrm{opt},m}^{\mathrm{H}}\mathbf{b}_k\right|^2 - 1\right)^2 \\ &= \sum_{m=1}^{N_{\mathrm{t}}} \left(\mathbf{b}_k^{\mathrm{H}}\mathbf{f}_{\mathrm{opt},m}\mathbf{f}_{\mathrm{opt},m}^{\mathrm{H}}\mathbf{b}_k - 1\right)^2\end{aligned}$$

$$= \left\| \left[\mathbf{b}_k^{\mathrm{H}} \left(\mathbf{f}_{\mathrm{opt},1}\mathbf{f}_{\mathrm{opt},1}^{\mathrm{H}} - \mathbf{I}_{n_t} \right) \mathbf{b}_k, \ldots, \mathbf{b}_k^{\mathrm{H}} \left(\mathbf{f}_{\mathrm{opt},N_t}\mathbf{f}_{\mathrm{opt},N_t}^{\mathrm{H}} - \mathbf{I}_{n_t} \right) \mathbf{b}_k \right] \right\|^2$$

$$= \left\| \left[\mathrm{vec}\left(\mathbf{f}_{\mathrm{opt},1}\mathbf{f}_{\mathrm{opt},1}^{\mathrm{H}} - \mathbf{I}_{n_t} \right), \ldots, \mathrm{vec}\left(\mathbf{f}_{\mathrm{opt},N_t}\mathbf{f}_{\mathrm{opt},N_t}^{\mathrm{H}} - \mathbf{I}_{n_t} \right) \right]^{\mathrm{H}} \mathrm{vec}\left(\mathbf{b}_k\mathbf{b}_k^{\mathrm{H}} \right) \right\|^2,$$

where the last step results from the identities $\mathrm{vec}(\mathbf{ABC}) = (\mathbf{C}^{\mathrm{T}} \otimes \mathbf{A})\mathrm{vec}(\mathbf{B})$ and $\mathrm{vec}(\mathbf{ab}^{\mathrm{T}}) = \mathbf{b} \otimes \mathbf{a}$. Note that $\mathbf{f}_{\mathrm{opt},i}\mathbf{f}_{\mathrm{opt},i}^{\mathrm{H}} - \mathbf{I}_{n_t}$, $1 \le i \le N_t$ and $\mathbf{b}_k\mathbf{b}_k^{\mathrm{H}}$ are Hermitian. For reasons that will become clear in the next section, let us define $\widetilde{\mathrm{vec}}(\cdot)$ for a Hermitian matrix, say $\mathbf{A}$, as

$$\widetilde{\mathrm{vec}}(\mathbf{A}) \triangleq \left[\mathrm{diag}(\mathbf{A})^{\mathrm{T}}, \sqrt{2}\Re\mathrm{vec}_1^{\mathrm{T}}(\mathbf{A}), \sqrt{2}\Im\mathrm{vec}_1^{\mathrm{T}}(\mathbf{A}) \right]^{\mathrm{T}},$$

and rewrite $\mathcal{L}_{k,\mathrm{ub}}(\mathbf{b}_k)$ equivalently as

$$\mathcal{L}_k^{\mathrm{ub}}(\mathbf{b}_k) = \Bigg\| \underbrace{\left[\widetilde{\mathrm{vec}}\left(\mathbf{f}_{\mathrm{opt},1}\mathbf{f}_{\mathrm{opt},1}^{\mathrm{H}} - \mathbf{I}_{n_t} \right), \ldots, \widetilde{\mathrm{vec}}\left(\mathbf{f}_{\mathrm{opt},N_t}\mathbf{f}_{\mathrm{opt},N_t}^{\mathrm{H}} - \mathbf{I}_{n_t} \right) \right]^{\mathrm{T}}}_{\triangleq \widetilde{\mathbf{G}}^{\mathrm{T}}} \widetilde{\mathrm{vec}}\left(\mathbf{b}_k\mathbf{b}_k^{\mathrm{H}} \right) \Bigg\|^2.$$

Therefore, the upper bound for $\mathcal{J}_{\mathrm{Tx}}$ becomes

$$\begin{aligned} \mathcal{J}_{\mathrm{Tx}}(\mathbf{B}_F) &\le \sum_{k=1}^{n_t} \mathcal{L}_k^{\mathrm{ub}}(\mathbf{b}_k) \\ &= \sum_{k=1}^{n_t} \widetilde{\mathrm{vec}}^{\mathrm{T}}\left(\mathbf{b}_k\mathbf{b}_k^{\mathrm{H}} \right) \widetilde{\mathbf{G}}\widetilde{\mathbf{G}}^{\mathrm{T}} \widetilde{\mathrm{vec}}\left(\mathbf{b}_k\mathbf{b}_k^{\mathrm{H}} \right) \\ &= \underbrace{\mathrm{tr}\left(\mathbf{Y}^{\mathrm{T}}\widetilde{\mathbf{G}}\widetilde{\mathbf{G}}^{\mathrm{T}}\mathbf{Y} \right)}_{\triangleq \mathcal{J}_{\mathrm{Tx}}^{\mathrm{ub}}(\mathbf{Y})}, \end{aligned}$$

where $\mathbf{Y} \triangleq [\widetilde{\mathrm{vec}}(\mathbf{b}_1\mathbf{b}_1^{\mathrm{H}}), \ldots, \widetilde{\mathrm{vec}}(\mathbf{b}_{n_t}\mathbf{b}_{n_t}^{\mathrm{H}})] \in \mathbb{R}^{n_t^2 \times n_t}$. The following theorem establishes the minimizing solution $\mathbf{Y}_{\mathrm{opt}}$ for the upper bound $\mathcal{J}_{\mathrm{Tx}}^{\mathrm{ub}}$.

Theorem 3.3 *The optimal* $\mathbf{Y}_{\mathrm{opt}} \in \mathbb{R}^{n_t^2 \times n_t}$ *that minimizes the upper bound* $\mathcal{J}_{\mathrm{Tx}}^{\mathrm{ub}}(\mathbf{Y}) = \mathrm{tr}(\mathbf{Y}^{\mathrm{T}}\widetilde{\mathbf{G}}\widetilde{\mathbf{G}}^{\mathrm{T}}\mathbf{Y})$ *is a subspace spanned by the eigenvectors of* $\widetilde{\mathbf{G}}\widetilde{\mathbf{G}}^{\mathrm{T}}$ *associated with the* n_t *smallest eigenvalues.*

Proof See Appendix 4.

Although a particular optimal $\mathbf{Y}_{\mathrm{opt}}$ is readily obtainable as the eigenvectors of $\widetilde{\mathbf{G}}\widetilde{\mathbf{G}}^{\mathrm{T}}$, it is the $\mathbf{b}_k$'s that are of interest to us. Observing that the solution is actually a subspace, we formulate retrieval of $\mathbf{b}_k$'s as joint diagonalization of matrices in the next section.

3.4.2 AJD-Based Solution

Let us express the Hermitian matrix $\widetilde{\mathbf{G}}\widetilde{\mathbf{G}}^{\mathrm{T}}$ by EVD as

$$\widetilde{\mathbf{G}}\widetilde{\mathbf{G}}^{\mathrm{T}} = [\mathbf{U}_{\widetilde{G},\perp}, \mathbf{U}_{\widetilde{G}}]\boldsymbol{\Lambda}_{\widetilde{G}}[\mathbf{U}_{\widetilde{G},\perp}, \mathbf{U}_{\widetilde{G}}]^{\mathrm{T}},$$

where $\boldsymbol{\Lambda}_{\widetilde{G}}$ is diagonal with eigenvalues in non-increasing order, and $\mathbf{U}_{\widetilde{G}} = [\mathbf{u}_{\widetilde{G},1}, \ldots, \mathbf{u}_{\widetilde{G},n_{\mathrm{t}}}] \in \mathbb{R}^{n_{\mathrm{t}}^2 \times n_{\mathrm{t}}}$. Following the conclusion of Theorem 3.3, the kth column of $\mathbf{Y}_{\mathrm{opt}}$, $\mathrm{vec}(\mathbf{b}_{\mathrm{opt},k}\mathbf{b}_{\mathrm{opt},k}^{\mathrm{H}})$, can be expressed by

$$\widetilde{\mathrm{vec}}\left(\mathbf{b}_{\mathrm{opt},k}\mathbf{b}_{\mathrm{opt},k}^{\mathrm{H}}\right) = \sum_{i=1}^{n_{\mathrm{t}}} \alpha_{k,i}\mathbf{u}_{\widetilde{G},i},\ 1 \le k \le n_{\mathrm{t}},$$

or in matrix terms,

$$\left[\widetilde{\mathrm{vec}}\left(\mathbf{b}_{\mathrm{opt},1}\mathbf{b}_{\mathrm{opt},1}^{\mathrm{H}}\right), \ldots, \widetilde{\mathrm{vec}}\left(\mathbf{b}_{\mathrm{opt},n_{\mathrm{t}}}\mathbf{b}_{\mathrm{opt},n_{\mathrm{t}}}^{\mathrm{H}}\right)\right] = [\mathbf{u}_{\widetilde{G},1}, \ldots, \mathbf{u}_{\widetilde{G},n_{\mathrm{t}}}]\boldsymbol{\alpha}, \tag{3.25}$$

where $\boldsymbol{\alpha} = [\alpha_{k,i}] \in \mathbb{C}^{n_{\mathrm{t}} \times n_{\mathrm{t}}}$ is the coefficient matrix. Because $\boldsymbol{\alpha}$ is non-singular as guaranteed by the linear independence of $\{\widetilde{\mathrm{vec}}(\mathbf{b}_{\mathrm{opt},k}\mathbf{b}_{\mathrm{opt},k}^{\mathrm{H}})\}_{k=1}^{n_{\mathrm{t}}}$, (3.25) is equivalent to

$$[\mathbf{u}_{\widetilde{G},1}, \ldots, \mathbf{u}_{\widetilde{G},n_{\mathrm{t}}}] = \left[\widetilde{\mathrm{vec}}\left(\mathbf{b}_{\mathrm{opt},1}\mathbf{b}_{\mathrm{opt},1}^{\mathrm{H}}\right), \ldots, \widetilde{\mathrm{vec}}\left(\mathbf{b}_{\mathrm{opt},n_{\mathrm{t}}}\mathbf{b}_{\mathrm{opt},n_{\mathrm{t}}}^{\mathrm{H}}\right)\right]\boldsymbol{\alpha}^{-1}.$$

Denote $\boldsymbol{\alpha}^{-1} = [\beta_{k,i}] \in \mathbb{C}^{n_{\mathrm{t}} \times n_{\mathrm{t}}}$ and $\boldsymbol{\Lambda}_k = \mathrm{diag}\{\beta_{k,1}, \ldots, \beta_{k,n_{\mathrm{t}}}\}$. Then for $1 \le k \le n_{\mathrm{t}}$,

$$\begin{aligned}
\mathbf{u}_{\widetilde{G},k} &= \sum_{i=1}^{n_{\mathrm{t}}} \beta_{k,i}\widetilde{\mathrm{vec}}\left(\mathbf{b}_{\mathrm{opt},i}\mathbf{b}_{\mathrm{opt},i}^{\mathrm{H}}\right) \\
&= \widetilde{\mathrm{vec}}\left(\sum_{i=1}^{n_{\mathrm{t}}} \mathbf{b}_{\mathrm{opt},i}\beta_{k,i}\mathbf{b}_{\mathrm{opt},i}^{\mathrm{H}}\right) \\
&= \widetilde{\mathrm{vec}}\left(\mathbf{B}_{\mathrm{opt}}\boldsymbol{\Lambda}_k\mathbf{B}_{\mathrm{opt}}^{\mathrm{H}}\right),
\end{aligned}$$

where $\mathbf{B}_{\mathrm{opt}} \triangleq [\mathbf{b}_{\mathrm{opt},1}, \ldots, \mathbf{b}_{\mathrm{opt},n_{\mathrm{t}}}]$. By defining $\mathbf{M}_{\widetilde{G},k}$ through $\mathbf{u}_{\widetilde{G},k} = \widetilde{\mathrm{vec}}(\mathbf{M}_{\widetilde{G},k})$, we finally arrive at

$$\widetilde{\mathrm{vec}}(\mathbf{M}_{\widetilde{G},k}) = \widetilde{\mathrm{vec}}\left(\mathbf{B}_{\mathrm{opt}}\boldsymbol{\Lambda}_k\mathbf{B}_{\mathrm{opt}}^{\mathrm{H}}\right),\ 1 \le k \le n_{\mathrm{t}}.$$

In other words,

$$\mathbf{M}_{\widetilde{G},k} = \mathbf{B}_{\mathrm{opt}}\boldsymbol{\Lambda}_k\mathbf{B}_{\mathrm{opt}}^{\mathrm{H}},\ 1 \le k \le n_{\mathrm{t}}.$$

Therefore, retrieval of $\mathbf{B}_{\text{opt}}$ from $\mathbf{Y}_{\text{opt}}$ is accomplished by jointly diagonalizing $\mathbf{M}_{\widetilde{G},k}$, $1 \le k \le n_t$. Note that from the construction of $\widetilde{\text{vec}}(\cdot)$, $\mathbf{M}_{\widetilde{G},k}$'s are Hermitian. Interestingly, such diagonalizing matrix can be found via low-complexity and numerically stable Jacobi-type algorithms [8, 20].

It is worth noting that because $\mathbf{M}_{\widetilde{G},k}$'s are generally not normal and commuting, such joint diagonalization is approximate in the sense that the off-diagonal entries of $\boldsymbol{\Lambda}_k$, denoted by off($\boldsymbol{\Lambda}_k$), can only be made arbitrarily small, i.e., $\sum_{k=1}^{n_t} \|\text{off}(\boldsymbol{\Lambda}_k)\|_F^2 \le \epsilon$ for $\epsilon > 0$ [20]. Thus, further normalization is required for $\mathbf{F}_{\text{RF}}^{\text{cm}} \in \mathcal{F}_{\text{RF}}$, i.e.,

$$\left[\mathbf{F}_{\text{RF}}^{\text{cm}}\right]_{i,k} = \frac{1}{\sqrt{N_t}} e^{\jmath \angle\left([\mathbf{F}_{\text{RF}}^{\text{opt}}\mathbf{B}_{\text{opt}}]_{i,k}\right)}, \; 1 \le i \le N_t, \; 1 \le k \le n_t.$$

The low-complexity procedure for constant-modulus RF design is summarized in Algorithm 3.1.

Algorithm 3.1 Constant-modulus RF analog precoder design via AJD

Input: Optimal design $\mathbf{F}_{\text{RF}}^{\text{opt}} = [\mathbf{f}_{\text{opt},1}, \ldots, \mathbf{f}_{\text{opt},N_t}]^{\text{H}} \in \mathbb{C}^{N_t \times n_t}$, AJD accuracy ϵ.
Output: Constant-modulus RF design $\mathbf{F}_{\text{RF}}^{\text{cm}} = e^{\jmath \angle(\mathbf{F}_{\text{RF}}^{\text{opt}}\mathbf{B}_{\text{opt}})}$.
$\widetilde{\mathbf{G}}$ = Empty matrix;
for $k = 1$ to N_t **do**
 $\widetilde{\mathbf{G}} = [\widetilde{\mathbf{G}} | \widetilde{\text{vec}}(\mathbf{f}_{\text{opt},k}\mathbf{f}_{\text{opt},k}^{\text{H}} - \mathbf{I}_{n_t})]$;
end for
Compute EVD $\widetilde{\mathbf{G}}\widetilde{\mathbf{G}}^{\text{T}} = [\mathbf{U}_{\widetilde{G},\perp}, \mathbf{u}_{\widetilde{G},1}, \ldots, \mathbf{u}_{\widetilde{G},n_t}]\boldsymbol{\Lambda}_{\widetilde{G}}[\mathbf{U}_{\widetilde{G},\perp}, \mathbf{u}_{\widetilde{G},1}, \ldots, \mathbf{u}_{\widetilde{G},n_t}]^{\text{T}}$;
for $k = 1$ to n_t **do**
 Reshape $\mathbf{u}_{\widetilde{G},k}$ such that $\widetilde{\text{vec}}(\mathbf{u}_{\widetilde{G},k}) = \mathbf{M}_{\widetilde{G},k}$;
end for
Obtain $\mathbf{B}_{\text{opt}}$ by repeated AJD of $\{\mathbf{M}_{\widetilde{G},k}\}_{k=1}^{n_t}$ such that $\sum_{k=1}^{n_t} \|\text{off}(\mathbf{B}_{\text{opt}}^{\text{H}}\mathbf{M}_{\widetilde{G},k}\mathbf{B}_{\text{opt}})\|_F^2 \le \epsilon$;

Overall, the proposed approach for two-timescale hybrid precoding and combining is formally stated in Algorithm 3.2.

Algorithm 3.2 Two-timescale hybrid precoding and MMSE combining

---

Statistical CSI-based RF design
Input: Transmit correlated eigen-space $\mathbf{U}_t$, receive correlated eigen-space $\mathbf{U}_r$, CCM $\boldsymbol{\Omega}$.
Output: $\mathbf{F}_{\text{RF}}^{\text{opt}} = \mathbf{U}_t\mathbf{S}_F^{\text{opt}}$, $\mathbf{W}_{\text{RF}}^{\text{opt}} = \mathbf{U}_r\mathbf{S}_W^{\text{opt}}$ (resp. $\mathbf{F}_{\text{RF}}^{\text{cm}}$, $\mathbf{W}_{\text{RF}}^{\text{cm}}$).
Obtain $\mathbf{Z}^{\text{opt}}$ by solving (3.19), and consequently $\mathbf{S}_F^{\text{opt}}$ and $\mathbf{S}_W^{\text{opt}}$ according to (3.20);
Obtain $\mathbf{F}_{\text{RF}}^{\text{cm}}$ and $\mathbf{W}_{\text{RF}}^{\text{cm}}$ by Algorithm 3.1 for constant-modulus design;
Instantaneous effective CSI-based baseband design
Input: CSI $\mathbf{H}_{\text{eff}} = (\mathbf{W}_{\text{RF}}^{\text{opt}})^{\text{H}}\mathbf{H}\mathbf{F}_{\text{RF}}^{\text{opt}}$ (resp. $\mathbf{H}_{\text{eff}} = (\mathbf{W}_{\text{RF}}^{\text{cm}})^{\text{H}}\mathbf{H}\mathbf{F}_{\text{RF}}^{\text{cm}}$), noise covariance $\mathbf{R}_n$.
Output: $\mathbf{F}_{\text{BB}}^{\text{opt}} = \mathbf{U}_{\text{BB}}^{\text{opt}}\boldsymbol{\Sigma}_{\text{BB}}^{\text{opt}}$, $\mathbf{W}_{\text{BB}}^{\text{opt}} = (\mathbf{H}_{\text{eff}}\mathbf{F}_{\text{BB}}^{\text{opt}}(\mathbf{F}_{\text{BB}}^{\text{opt}})^{\text{H}}\mathbf{H}_{\text{eff}}^{\text{H}} + \mathbf{I}_{n_r})^{-1}\mathbf{H}_{\text{eff}}\mathbf{F}_{\text{BB}}^{\text{opt}}$.
Compute EVD $\mathbf{H}_{\text{eff}}^{\text{H}}\mathbf{R}_n^{-1}\mathbf{H}_{\text{eff}} = \mathbf{U}_{\mathbf{R}_H}\boldsymbol{\Lambda}_{\mathbf{R}_H}\mathbf{V}_{\mathbf{R}_H}^{\text{H}}$;
Baseband beamformer $\mathbf{U}_{\text{BB}}^{\text{opt}} = [\mathbf{V}_{\mathbf{R}_H}]_{1,\ldots,N_s}^{1,\ldots,n_t}$;
Power allocation $\boldsymbol{\Sigma}_{\text{BB}}^{\text{opt}}$ by waterfilling (3.11), (3.12);

Remark 3.3 In Algorithm 3.1, to obtain the constant-modulus RF combiner $\mathbf{W}_{\mathrm{RF}}^{\mathrm{cm}}$, we simply make the following substitution: $\mathbf{F}_{\mathrm{RF}}^{\mathrm{opt}} \leftarrow \mathbf{W}_{\mathrm{RF}}^{\mathrm{opt}}$, $n_{\mathrm{t}} \leftarrow n_{\mathrm{r}}$, and $N_{\mathrm{t}} \leftarrow N_{\mathrm{r}}$.

Remark 3.4 Existing methods such as sparse approximation [1] and alternating optimization [2] attempt to reconstruct the optimal design by combining the RF and baseband designs together. When two-timescale CSI is considered, such approaches are not directly applicable. By separating the hybrid design into two stages, our approach yields the optimal channel-diagonalizing precoder and LMMSE combiner at baseband while facilitating derivation of the statistical CSI-based constant-modulus RF solutions.

3.5 Illustrative Results and Discussions

In this section, we numerically evaluate the effective rates of the proposed two-timescale hybrid solutions under the impact of (1) a reduced number of RF chains, (2) loose and stringent delay-outage constraints, (3) varied angle spread.

3.5.1 *Simulation Setup*

Suppose that both link ends are equipped with uniform planar arrays of sector antennas. The double-directional MIMO channel is parametrically generated as

$$\mathbf{H} = \gamma \sum_{i=1}^{N_{\mathrm{cl}}} \sum_{l=1}^{N_{\mathrm{ray}}} \alpha_{il} g_{\mathrm{r}}\left(\phi_{il}^{\mathrm{r}}, \theta_{il}^{\mathrm{r}}\right) \mathbf{a}_{\mathrm{r}}\left(\phi_{il}^{\mathrm{r}}, \theta_{il}^{\mathrm{r}}\right) g_{\mathrm{t}}\left(\phi_{il}^{\mathrm{t}}, \theta_{il}^{\mathrm{t}}\right) \mathbf{a}_{\mathrm{t}}\left(\phi_{il}^{\mathrm{t}}, \theta_{il}^{\mathrm{t}}\right)^{\mathrm{H}}.$$

It is assumed that the transmitter and the receiver are coupled via N_{cl} clusters, each with N_{ray} scattering paths. For the lth ray in the ith cluster with an azimuth AoD ϕ_{il}^{t}, AoA ϕ_{il}^{r}, and an elevation AoD θ_{il}^{t}, AoA θ_{il}^{r}, α_{il} represents the path gain, $g_{\mathrm{t}}(\phi_{il}^{\mathrm{t}}, \theta_{il}^{\mathrm{t}})$ is the transmit antenna gain, $g_{\mathrm{r}}(\phi_{il}^{\mathrm{r}}, \theta_{il}^{\mathrm{r}})$ is the receive antenna gain, $\mathbf{a}_{\mathrm{t}}(\phi_{il}^{\mathrm{t}}, \theta_{il}^{\mathrm{t}})$ and $\mathbf{a}_{\mathrm{r}}(\phi_{il}^{\mathrm{r}}, \theta_{il}^{\mathrm{r}})$ are the transmit and receive array responses respectively. In particular, the transmit antenna gain is approximated by $g_{\mathrm{t}}(\phi_{il}^{\mathrm{t}}, \theta_{il}^{\mathrm{t}}) = g_{\mathrm{az}}^{\mathrm{t}}\left(\phi_{il}^{\mathrm{t}}\right) g_{\mathrm{el}}^{\mathrm{t}}\left(\theta_{il}^{\mathrm{t}}\right)$ [21, 22], where

$$g_{\mathrm{az}}^{\mathrm{t}}\left(\phi_{il}^{\mathrm{t}}\right) = -12\left(\frac{\phi_{il}^{\mathrm{t}}}{\phi_{3\mathrm{dB}}^{\mathrm{t}}}\right)^2 \mathrm{dB}, \quad g_{\mathrm{el}}^{\mathrm{t}}\left(\theta_{il}^{\mathrm{t}}\right) = -12\left(\frac{\theta_{il}^{\mathrm{t}}}{\theta_{3\mathrm{dB}}^{\mathrm{t}}}\right)^2 \mathrm{dB}$$

are the horizontal gain with a half-power beamwidth $\phi_{3\mathrm{dB}}^{\mathrm{t}}$ and the vertical gain with a half-power beamwidth $\theta_{3\mathrm{dB}}^{\mathrm{t}}$ respectively. Further, the transmit array response $\mathbf{a}_{\mathrm{t}}(\phi_{il}^{\mathrm{t}}, \theta_{il}^{\mathrm{t}})$ is defined as

$$\mathbf{a}_{\mathrm{t}}\left(\phi_{il}^{\mathrm{t}}, \theta_{il}^{\mathrm{t}}\right) = \Big[1, \ldots, e^{\jmath 2\pi d_{\lambda}\left(m \sin(\phi_{il}^{\mathrm{t}}) \sin(\theta_{il}^{\mathrm{t}}) + n \cos(\theta_{il}^{\mathrm{t}})\right)}, \ldots,$$
$$e^{\jmath 2\pi d_{\lambda}\left((Y_{\mathrm{t}}-1) \sin(\phi_{il}^{\mathrm{t}}) \sin(\theta_{il}^{\mathrm{t}}) + (Z_{\mathrm{t}}-1) \cos(\theta_{il}^{\mathrm{t}})\right)}\Big]^{\mathrm{T}},$$

where $0 \leq m \leq Y_t - 1$ and $0 \leq n \leq Z_t - 1$ are the antenna indices along the Y-dimension and Z-dimension, respectively, and $d_\lambda = 0.5$ for critically spaced antenna elements. The receive antenna gain $g_r(\phi^r_{il}, \theta^r_{il})$ and array response $\mathbf{a}_r(\phi^r_{il}, \theta^r_{il})$ are similarly defined. γ is the normalization factor such that $\mathbb{E}[\|\mathbf{H}\|_F^2] = N_r N_t$.

The channel parameters for simulations are set up as follows. There are $N_{cl} = 5$ clusters, each with $N_{ray} = 20$ scattering paths [23]. The path gains $\{\alpha_{il}\}$ are independent standard Gaussian random variables. The half-power beamwidths are chosen as $\phi^t_{3dB} = \phi^r_{3dB} = 35^\circ$ and $\theta^t_{il} = \theta^r_{il} = 15^\circ$. The truncated Laplacian distribution is used to generate the azimuth AoDs $\{\phi^t_{il}\}$, AoAs $\{\phi^r_{il}\}$ with the mean uniformly distributed over a range of 35° and the elevation AoDs $\{\theta^t_{il}\}$ and AoAs $\{\theta^r_{il}\}$ with the mean uniformly distributed over a range of 15°. A common angle spread is assumed. The transmit correlation matrix $\mathbf{R}_t = \mathbb{E}[\mathbf{H}^H\mathbf{H}]$ in (3.3), the receive correlation matrix $\mathbf{R}_r = \mathbb{E}[\mathbf{H}\mathbf{H}^H]$ in (3.4), and the CCM $\boldsymbol{\Omega}$ in (3.2) are obtained by Monte Carlo simulations over a total of 5000 channel realizations.

The effective rates of the unconstrained-modulus design (denoted by Eigenmode) as well as the constant-modulus design (denoted by AJD) are plotted versus the normalized nominal SNR $\rho = P_t$. The normalized fading block length and the maximum queue length are assumed as $BT = 500$ and $Q_{max} = 1500$ respectively. $N_s = 4$ data streams are transmitted with waterfilling-based power allocation. The effective rates of three *perfect CSI*-based benchmarks are also presented for comparison: (1) joint fully digital Tx-Rx design (denoted by Optimal) [12]; (2) joint hybrid Tx-Rx solution via orthogonal matching pursuit (denoted by OMP) [1]; (3) joint hybrid Tx-Rx solution with phase extraction for RF precoding and combining (denoted by PE) [24]. Note that schemes (2) and (3) assume constant-modulus RF elements.

3.5.2 *Effect of Limited RF Chains*

It is observed in Fig. 3.2 that subject to a loose delay-outage constraint represented by a delay-outage limit $\kappa = 0.95$, the performance gap is fairly small ($\approx 1\%$ loss) between the hybrid eigenmode solution using a very limited number of RF chains (relative to the number of antennas) and the fully digital optimal solution where a separate RF chain is attached to each antenna element. For a stringent delay-outage limit $\kappa = 10^{-3}$, the performance gap remains almost the same. Similar observations can be made in Fig. 3.3 when the number of transmit and receive antennas is doubled. Note that a small increase in the number of RF chains suffices for Eigenmode to perform close to the optimal solution. An intuitive explanation is that the channel coupling power tends to concentrate in a low-dimensional eigen-domain when directional scattering is observed, and therefore only a small performance loss results from reduction in the high dimension of the MIMO channel

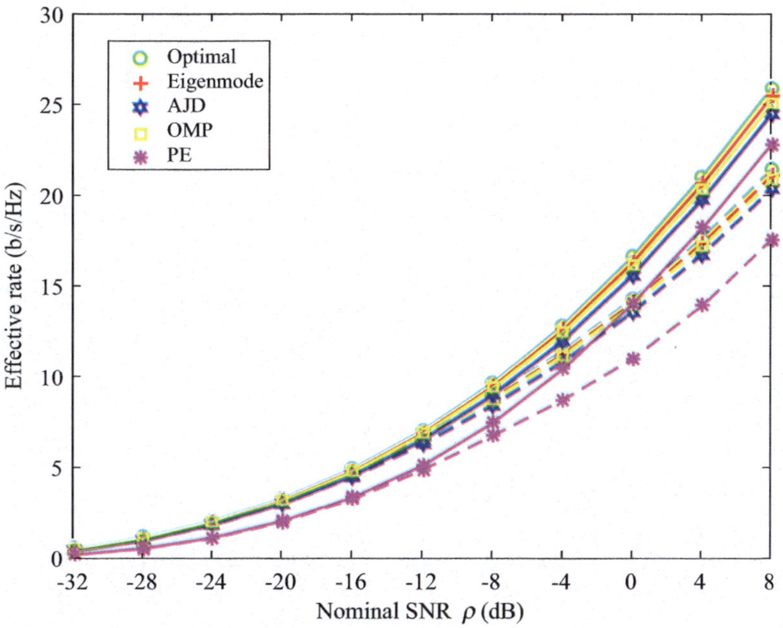

Fig. 3.2 Effective rates achieved by various joint Tx-Rx solutions for a 64 × 16 massive MIMO system under loose and stringent delay-outage constraints. $n_t = 8$ transmit and $n_r = 5$ receive RF chains are employed. The angle spread is 6°. Solid lines correspond to a delay-outage limit $\kappa = 0.95$ and dash lines to $\kappa = 10^{-3}$

by RF preprocessing. Although the joint eigenmode solutions are derived regardless of the normalized QoS exponent ξ for the sake of mathematical tractability, the penalty incurred by such an approximate approach nonetheless turns out to be negligible. Interestingly, Eigenmode even offers slightly better effective rates than OMP. Therefore, it is safe to conclude that in a directional scattering environment, two-timescale hybrid design may serve as a viable option which delivers a near-optimal effective rate under loose and stringent delay-outage constraints. Since it is error-prone to estimate instantaneous realizations of a high-dimensional channel, the two-timescale design is expected to even outperform the fully digital design when channel estimation errors are taken into account.

When constant-modulus RF elements are employed, our proposed two-timescale CSI-based solution performs only slightly worse ($\approx$2% loss for a 64 × 16 system and $\approx$3% loss for a 128 × 32 system) compared with the OMP-based counterpart which relies on perfect CSI at the transmitter and the receiver. On the other hand, our solution offers a significant performance gain over PE which attempts to achieve coherent signal superposition by negating the phase effect of the channel in the RF domain. The goodness of the proposed solution AJD arises from its closeness ($\approx$3% loss) of performance to the unconstrained-modulus counterpart. Because the effective rate is directly related to the eigenvalues of the low-dimensional effective

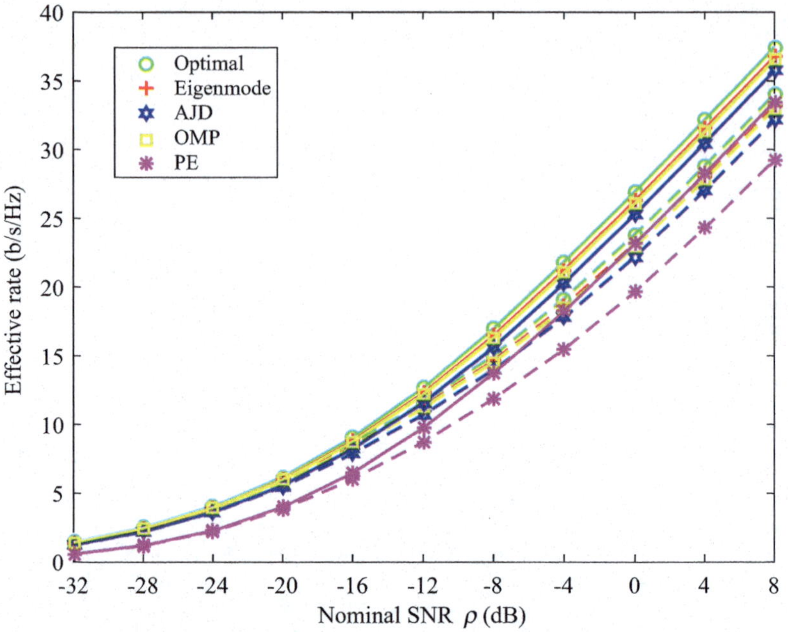

Fig. 3.3 Effective rates achieved by various joint Tx-Rx solutions for a 128×32 massive MIMO system under loose and stringent delay-outage constraints. $n_\mathrm{t} = 10$ transmit and $n_\mathrm{r} = 6$ receive RF chains are employed. The angle spread is 6°. Solid lines correspond to a delay-outage limit $\kappa = 0.95$ and dash lines to $\kappa = 10^{-3}$

channel, this nearness implies that the proposed magnitude LS method yields a good approximation for the eigen-structure of the eigenmode selection design, and therefore the good performance of AJD.

3.5.3 *Effect of Delay-Outage Constraints*

To better assess the impact of delay-outage constraints, we plot the effective rates of hybrid schemes together with the fully digital optimal design from loose to stringent delay-outage constraints in Fig. 3.4. Interestingly, the two-timescale solution offers robust optimality-achieving performance with respect to various delay-outage constraints. It is worth pointing out that by exploiting only statistical CSI in the RF domain, AJD performs fairly close to OMP while offering a noticeable performance gain over PE throughout the delay-outage limits under consideration. This is attributed to the fact that the proposed AJD-based approach arrives at a close approximation of the eigen-structure derived from the unconstrained-modulus counterpart, which is near optimal in this case. Since Eigenmode slightly outperforms OMP, the unconstrained-modulus design can be a useful alternative in scenarios where channel training overhead is a major concern.

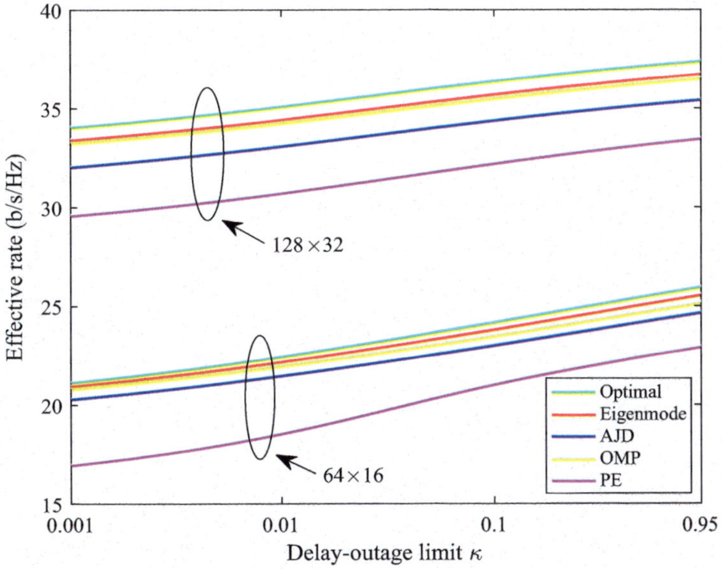

Fig. 3.4 Impact of delay-outage constraint on the effective rates achieved by various joint Tx-Rx solutions at a nominal SNR 8 dB. The angle spread is 6°. For a 128×32 massive MIMO system, $n_t = 10$ and $n_r = 6$ RF chains are employed and for a 64×16 system, $n_t = 8$ and $n_r = 5$ RF chains are employed

3.5.4 Effect of Angle Spread

Figure 3.5 illustrates the behavior of the fully digital and the hybrid designs in response to varying degrees of scattering under loose and stringent delay-outage constraints respectively. In general, an increased angle spread translates to an increased spatial diversity, which in turn enables the precoding and combining solutions to mitigate the effect of delay-outage constraint. This is indeed the case when the angle spread is small ($< 5^{\circ}$). It is seen that performance fluctuates as angle spread increases when a stringent delay-outage limit is imposed. The intuitive reason is that when the angle spread is no longer much smaller than the beamwidths of the sector antennas, the relation between angle spread and spatial diversity does not necessarily hold. As a result, the effective rate might be noticeably affected by the delay-outage limit, and thus manifests fluctuation. Such a phenomenon becomes less severe when larger antenna arrays (in this case, 128×32) are deployed. Finally, when the scattering becomes rich, i.e., a large angle spread around 15°, the channel power tends to spread out in the eigen-domain. When only a limited number of data streams are transmitted, the service rate suffers from decreased eigen-channel gains. This in turn negatively affects the arrival rate. Interestingly, due to directional scattering, the two-timescale Eigenmode performs close to the full-complexity optimal solutions irrespective of the angle spread. Moreover, since

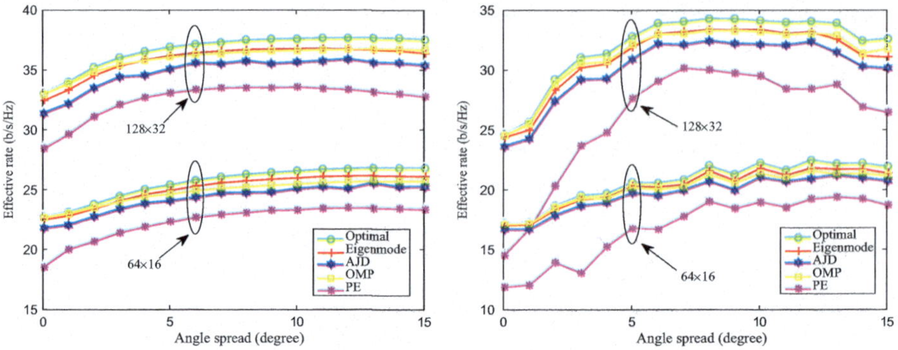

Fig. 3.5 Impact of angle spread on the effective rates achieved by various joint Tx-Rx solutions. The nominal SNR is 8 dB. For a 128 × 32 massive MIMO system, $n_\mathrm{t} = 10$ and $n_\mathrm{r} = 6$ RF chains are employed, and for a 64 × 16 system, $n_\mathrm{t} = 8$ and $n_\mathrm{r} = 5$ RF chains are employed. The left subfigure corresponds to a delay-outage limit $\kappa = 0.95$ and the right to $\kappa = 10^{-3}$

our proposed approach to constant-modulus design well approximates the eigen-structure of Eigenmode, it gives a good performance as long as Eigenmode remains near optimal, and therefore incurs only a small performance loss (<3%) compared with OMP.

3.5.5 Computational Complexity

Regardless of the cost of CSI acquisition, we evaluate the computational complexity in terms of real flops (addition and multiplication) for the presented schemes. To gain intuitive insights, only the dominating terms are kept, which are determined based on the simulation parameters for the 64 × 16 massive MIMO system. The full-complexity Optimal solutions generally require $8N_\mathrm{t}^2N_\mathrm{r} + \mathcal{O}(N_\mathrm{t}^3)$ for precoding and $8N_\mathrm{r}N_\mathrm{t}N_\mathrm{s} + \mathcal{O}(N_\mathrm{r}^3)$ for linear MMSE combining. With the knowledge of such optimal solutions, OMP computes the hybrid precoder at the complexity of $8N_\mathrm{cl}N_\mathrm{ray}N_\mathrm{t}n_\mathrm{t}N_\mathrm{s} + 8n_\mathrm{t}^2N_\mathrm{t}N_\mathrm{s} + 8n_\mathrm{t}^3N_\mathrm{t}$ and the hybrid combiner at $8N_\mathrm{cl}N_\mathrm{ray}N_\mathrm{r}n_\mathrm{r}N_\mathrm{s} + 8N_\mathrm{cl}N_\mathrm{ray}n_\mathrm{r}N_\mathrm{r}^2 + 8n_\mathrm{r}^2N_\mathrm{r}^2$. As the complexity of phase extraction of the optimal solutions in the RF domain is negligible, additional complexity incurred by PE is attributed to the derivation of the baseband solutions, which totally requires $8n_\mathrm{r}N_\mathrm{t}N_\mathrm{r} + 8n_\mathrm{t}n_\mathrm{r}N_\mathrm{s}$. Recall that Optimal, OMP, and PE assume perfect CSI knowledge, and therefore computation takes place per channel coherence period. For the proposed Eigenmode solution, the expense of RF design comes primarily from solving the SDP optimization problem, which typically needs $\mathcal{O}((N_\mathrm{t} + N_\mathrm{r})^3)$ in light of the sparse structure of $\boldsymbol{\Omega}$. Together with the baseband design, overall, it requires $8n_\mathrm{r}N_\mathrm{t}N_\mathrm{r} + 8n_\mathrm{r}n_\mathrm{t}N_\mathrm{t} + \mathcal{O}((N_\mathrm{t} + N_\mathrm{r})^3)$ for joint hybrid precoding and combining solutions. The constant-modulus precoding (and combining) solution by AJD involves EVD of $\widetilde{\mathbf{G}}\widetilde{\mathbf{G}}^\mathrm{T}$ at $n_\mathrm{t}^4N_\mathrm{t} + \mathcal{O}\left(n_\mathrm{t}^6\right)$ (and $n_\mathrm{r}^4N_\mathrm{r} + \mathcal{O}\left(n_\mathrm{r}^6\right)$), and joint matrix diagonalization [20] at $\mathcal{O}\left(n_t^3\right)$ (and $\mathcal{O}\left(n_r^3\right)$) per iteration. In combination with the baseband design, the total number of flops amounts to $8n_r N_t N_r + n_t^4 N_t + \mathcal{O}(n_t^6)$ for hybrid precoding and $8n_r^2 N_r + n_r^4 N_r + \mathcal{O}\left(n_r^6\right)$ for hybrid combining. For the two-step approaches such as Eigenmode and AJD, the computation for RF components accounts for a large portion of the overall computational cost. Although AJD suffers a slight complexity increase per update in comparison with OMP, fortunately, it suffices to update the RF components once per multiple channel coherence periods. This is further illustrated in Table 3.1, where the running time of various joint Tx-Rx solutions is measured in millisecond (ms) on the assumption that the channel statistics remain invariant over 100 channel fading blocks [25]. The measurements are carried out on a Dell XPS 8700 (Core Quad i7-4770 at 3.9 GHz with 16 GB of DDR3 RAM) with MATLAB 2016a.

Table 3.1 Comparison of computational complexity (running time)

Schemes	Optimal	Eigenmode	AJD	OMP	PE
Running time (ms)	366	171	387	615	39

3.6 Summary

In this chapter, we investigated a joint design of two-timescale hybrid precoding and combining for massive MIMO wireless backhaul, where the objective was to maximize the mutual information subject to the statistical queueing constraint. The conventional MIMO solutions, i.e., the optimal channel-diagonalizing precoding and linear MMSE combining were employed for the fully digital baseband processing which was adaptive to the instantaneous CSI. The optimization problem boiled down to solving for the analog transmit-receive solutions in the RF domain using the statistical CSI. The cases of nonuniform- and constant-modulus RF elements were studied. In the absence of a closed-form expression for the objective function, the optimal structures of the RF analog precoder and combiner were derived, which led to a combinatorial problem of eigenmode selection. The technique of SDR was exploited to obtain an approximate solution to the NP-hard problem. In order to handle the nonconvex constant-modulus constraint, a matrix reconstruction formulation was proposed and solved by low-complexity Jacobi-like algorithms. Through numerical simulations, we demonstrated the effectiveness of the two-timescale hybrid RF-baseband design in terms of its near-optimal performance under both loose and stringent delay-outage constraints when directional scattering was present. Besides, the proposed constant-modulus solution was shown to well approximate the eigen-structure of the nonuniform counterpart, and offered comparable effective rates with other perfect CSI-based state of the arts.

Appendix 1: Proof of Theorem 3.1

Recall that in the general case $\mathbf{H} = \mathbf{U}_{\mathrm{r}}\tilde{\mathbf{H}}\mathbf{U}_{\mathrm{t}}^{\mathrm{H}}$, and accordingly

$$\begin{aligned}\mathbf{R}_{\mathrm{eff}} &= \mathbf{F}_{\mathrm{RF}}^{\mathrm{H}}\mathbf{H}^{\mathrm{H}}\mathbf{W}_{\mathrm{RF}}\mathbf{W}_{\mathrm{RF}}^{\mathrm{H}}\mathbf{H}\mathbf{F}_{\mathrm{RF}} \\ &= \mathbf{F}_{\mathrm{RF}}^{\mathrm{H}}\mathbf{U}_{\mathrm{t}}\tilde{\mathbf{H}}^{\mathrm{H}}\underbrace{\mathbf{U}_{\mathrm{r}}^{\mathrm{H}}\mathbf{W}_{\mathrm{RF}}\mathbf{W}_{\mathrm{RF}}^{\mathrm{H}}\mathbf{U}_{\mathrm{r}}}_{\triangleq \boldsymbol{\Xi}}\tilde{\mathbf{H}}\mathbf{U}_{\mathrm{t}}^{\mathrm{H}}\mathbf{F}_{\mathrm{RF}}.\end{aligned}$$

To show that the optimal $\mathbf{W}_{\mathrm{RF}}^{\mathrm{opt}}$ can be expressed as a subset from the receive eigenmodes, we follow a similar line of reasoning as in [26]. Recall that by substituting in the optimal baseband precoder $\mathbf{F}_{\mathrm{BB}}^{\mathrm{opt}} = \mathbf{V}_F\boldsymbol{\Sigma}_F^{-1}\mathbf{U}_{\mathbf{R}_{\mathrm{eff}}}\boldsymbol{\Sigma}_{\mathrm{BB}}^{\mathrm{opt}}$, the objective function becomes

$$C(\boldsymbol{\Xi}) = -\frac{1}{\xi}\log_2\mathbb{E}\left[\prod_{i=1}^{N_s}\left(1+\sigma_i^2\lambda_{[i],\mathbf{R}_{\mathrm{eff}}}\right)^{-\xi}\right].$$

It has been shown in [6] that C is concave in $\boldsymbol{\Xi}$. The idea is to argue that C is maximized only when $\boldsymbol{\Xi}$ is diagonal. To start with, let $\boldsymbol{\Pi}_i$ denote a matrix equal to the identity matrix except for the ith diagonal element being -1. Accordingly, a new matrix $\tilde{\boldsymbol{\Xi}}_i$, which is obtained from $\boldsymbol{\Xi}$ by setting to zero its ith row and ith column except the ith diagonal element, can be expressed as

$$\tilde{\boldsymbol{\Xi}}_i = \frac{1}{2}\left(\boldsymbol{\Xi} + \boldsymbol{\Pi}_i^{\mathrm{H}}\boldsymbol{\Xi}\boldsymbol{\Pi}_i\right).$$

Using Jensen's inequality, we conclude

$$C\left(\tilde{\boldsymbol{\Xi}}_i\right) \geq \frac{1}{2}C(\boldsymbol{\Xi}) + \frac{1}{2}C\left(\boldsymbol{\Pi}_i^{\mathrm{H}}\boldsymbol{\Xi}\boldsymbol{\Pi}_i\right).$$

Further, since the elements of $\tilde{\mathbf{H}}$ are zero mean and independent, changing the signs of its elements does not change the distribution, i.e., $\boldsymbol{\Pi}_i\tilde{\mathbf{H}}$ has the same distribution as $\tilde{\mathbf{H}}$. Thus, by direct substitution, it can be shown that

$$C\left(\boldsymbol{\Pi}_i^{\mathrm{H}}\boldsymbol{\Xi}\boldsymbol{\Pi}_i\right) = C(\boldsymbol{\Xi}),$$

which translates to

$$C\left(\tilde{\boldsymbol{\Xi}}_i\right) \geq C(\boldsymbol{\Xi}).$$

By iterating this process, we argue that the optimal $\boldsymbol{\Xi}$ must be diagonal. In that $\mathbf{U}_{\mathrm{W}}$ is semi-unitary, this can be achieved when

$$\mathbf{W}_{\mathrm{RF}} = \mathbf{U}_{\mathrm{r}}\mathbf{S}_W.$$

Using the fact that **AB** and **BA** share the same non-zero eigenvalues, we see that

$$\lambda_{[i],\mathbf{R}_{\text{eff}}} = \lambda_{[i]}\left(\mathbf{W}_{\text{RF}}^{\text{H}}\mathbf{H}\mathbf{F}_{\text{RF}}\mathbf{F}_{\text{RF}}^{\text{H}}\mathbf{H}^{\text{H}}\mathbf{W}_{\text{RF}}\right),$$

and thus the conclusion that the optimal $\mathbf{F}_{\text{RF}}$ selects n_{t} out of N_{t} transmit eigenmodes can be similarly proved.

Appendix 2: Proof of Corollary 3.1

In the case of separable correlation, $\mathbf{H} = \mathbf{U}_{\text{r}}\boldsymbol{\Lambda}_{\text{r}}\mathbf{H}_{\text{iid}}\boldsymbol{\Lambda}_{\text{t}}\mathbf{U}_{\text{t}}^{\text{H}}$ where $\boldsymbol{\Lambda}_{\text{r}}$ and $\boldsymbol{\Lambda}_{\text{t}}$ have eigenvalues on the diagonal in non-increasing order. Following the conclusion of Theorem 3.1, we have

$$\begin{aligned}\mathbf{R}_{\text{eff}} &= \mathbf{F}_{\text{RF}}^{\text{H}}\mathbf{H}^{\text{H}}\mathbf{W}_{\text{RF}}\mathbf{W}_{\text{RF}}^{\text{H}}\mathbf{H}\mathbf{F}_{\text{RF}} \\ &= \mathbf{S}_{F}^{\text{T}}\boldsymbol{\Lambda}_{\text{t}}\underbrace{\mathbf{H}_{\text{iid}}^{\text{H}}\boldsymbol{\Lambda}_{\text{r}}\mathbf{S}_{W}\mathbf{S}_{W}^{\text{T}}\boldsymbol{\Lambda}_{\text{r}}}_{\triangleq\boldsymbol{\Psi}(\mathbf{S}_W)}\mathbf{H}_{\text{iid}}\boldsymbol{\Lambda}_{\text{t}}\mathbf{S}_{F}.\end{aligned} \tag{3.26}$$

Let $\mathbf{S}_{W}^{\text{opt}}$ denote the selection matrix that picks the n_{r} largest eigenvalues out of $\boldsymbol{\Lambda}_{\text{r}}$ and $\mathbf{S}_{W}^{\text{alt}}$ represent any other choice. Obviously,

$$\boldsymbol{\Psi}\left(\mathbf{S}_{W}^{\text{opt}}\right) \succeq \mathbf{P}^{\text{T}}\boldsymbol{\Psi}\left(\mathbf{S}_{W}^{\text{alt}}\right)\mathbf{P}$$

where the permutation matrix $\mathbf{P}$ rearranges the diagonal elements of $\boldsymbol{\Psi}(\mathbf{S}_{W}^{\text{alt}})$ in non-decreasing order. Using Weyl's inequality [13], we conclude that

$$\lambda_{\mathbf{R}_{\text{eff}}^{\text{opt}},[i]} \geq \lambda_{\mathbf{R}_{\text{eff}}^{\text{alt}},[i]},\ 1 \leq i \leq n_{\text{r}},$$

where $\mathbf{R}_{\text{eff}}^{\text{opt}}$ and $\mathbf{R}_{\text{eff}}^{\text{alt}}$ are obtained by replacing $\boldsymbol{\Psi}(\mathbf{S}_W)$ with $\boldsymbol{\Psi}(\mathbf{S}_{W}^{\text{opt}})$ and $\mathbf{P}^{\text{T}}\boldsymbol{\Psi}(\mathbf{S}_{W}^{\text{alt}})\mathbf{P}$ in (3.26) respectively. Translating to the effective rate, it holds that

$$\begin{aligned}-\frac{1}{\xi}\log_2\mathbb{E}_{\mathbf{H}}\left[\prod_{i=1}^{N_{\text{s}}}\left(1+\sigma_i^2\lambda_{\mathbf{R}_{\text{eff}}^{\text{opt}},[i]}\right)^{-\xi}\right] &\geq -\frac{1}{\xi}\log_2\mathbb{E}_{\mathbf{H}}\left[\prod_{i=1}^{N_{\text{s}}}\left(1+\sigma_i^2\lambda_{\mathbf{R}_{\text{eff}}^{\text{alt}},[i]}\right)^{-\xi}\right] \\ &= -\frac{1}{\xi}\log_2\mathbb{E}_{\mathbf{H}}\left[\prod_{i=1}^{N_{\text{s}}}\left(1+\sigma_i^2\lambda_{[i]}\left(\mathbf{S}_{F}^{\text{T}}\boldsymbol{\Lambda}_{\text{t}}\mathbf{H}_{\text{iid}}^{\text{H}}\boldsymbol{\Psi}\left(\mathbf{S}_{W}^{\text{alt}}\right)\mathbf{H}_{\text{iid}}\boldsymbol{\Lambda}_{\text{t}}\mathbf{S}_{F}\right)\right)^{-\xi}\right],\end{aligned}$$

where the last step comes from the fact that since the elements of $\mathbf{H}_{\text{iid}}$ are i.i.d., $\mathbf{P}\mathbf{H}_{\text{iid}}$ has the same distribution as $\mathbf{H}_{\text{iid}}$. The case of $\mathbf{S}_F$ can be similarly proved.

Appendix 3: Proof of Theorem 3.2

Recall that $\mathbf{R}_{\text{eff}} = \mathbf{S}_F^{\text{H}}\tilde{\mathbf{H}}^{\text{H}}\mathbf{S}_W\mathbf{S}_W^{\text{H}}\tilde{\mathbf{H}}\mathbf{S}_F$ as a result of Theorem 3.1. Let $d_k = [\mathbf{R}_{\text{eff}}]_{k,k}$ the kth diagonal element of $\mathbf{R}_{\text{eff}}$. It is straightforward to show that its expectation is

$$\mathbb{E}[d_k] = \sum_{l=1}^{n_{\text{r}}} [\boldsymbol{\Omega}]_{i_l,j_k}.$$

To derive the upper bound, we first observe the following approximation to the effective rate, i.e.,

$$\begin{aligned}
\mathcal{C} &= -\frac{1}{\xi}\log_2 \mathbb{E}\left[\prod_{k=1}^{N_{\text{s}}}\left(1+\sigma_k^2\lambda_{\mathbf{R}_{\text{eff}},[k]}\right)^{-\xi}\right] \\
&\approx -\frac{1}{\xi}\log_2 \mathbb{E}\left[\prod_{k=1}^{N_{\text{s}}}\left(1+\frac{\rho}{N_{\text{s}}}\lambda_{\mathbf{R}_{\text{eff}},[k]}\right)^{-\xi}\right] \\
&\approx N_{\text{s}}\log_2\frac{\rho}{N_{\text{s}}} - \frac{1}{\xi}\log_2 \mathbb{E}\left[\left(\prod_{k=1}^{N_{\text{s}}}\lambda_{\mathbf{R}_{\text{eff}},[k]}\right)^{-\xi}\right],
\end{aligned}$$

which results from the fact that in the high effective SNR regime, i.e., $\lambda_{\mathbf{R}_{\text{eff}},[k]} \gg 1$, $1 \leq k \leq L$, as enabled by the high beamforming gain produced by large-scale antenna arrays [1], equal power allocation is asymptotically optimal. Furthermore, we have

$$\begin{aligned}
& N_{\text{s}}\log_2\frac{\rho}{N_{\text{s}}} - \frac{1}{\xi}\log_2 \mathbb{E}\left[\left(\prod_{k=1}^{N_{\text{s}}}\lambda_{\mathbf{R}_{\text{eff}},[k]}\right)^{-\xi}\right] \\
&\overset{(a)}{\leq} N_{\text{s}}\log_2\frac{\rho}{N_{\text{s}}} - \frac{1}{\xi}\log_2 \mathbb{E}\left[\left(\prod_{k=1}^{n_{\text{t}}} d_k\right)^{-\xi}\right] \\
&\overset{(b)}{\leq} N_{\text{s}}\log_2\frac{\rho}{N_{\text{s}}} + n_{\text{t}}\log_2\left(\prod_{k=1}^{n_{\text{t}}}\mathbb{E}[d_k]\right)^{\frac{1}{n_{\text{t}}}} \\
&\overset{(c)}{\leq} N_{\text{s}}\log_2\frac{\rho}{N_{\text{s}}} + n_{\text{t}}\log_2\left(\frac{1}{n_{\text{t}}}\sum_{k=1}^{n_{\text{t}}}\mathbb{E}[d_k]\right) \\
&= N_{\text{s}}\log_2\frac{\rho}{N_{\text{s}}} - n_{\text{t}}\log_2 n_{\text{t}} + n_{\text{t}}\log_2\sum_{l=1}^{n_{\text{r}}}\sum_{k=1}^{n_{\text{t}}}[\boldsymbol{\Omega}]_{i_l,j_k},
\end{aligned}$$

where Hadamard's inequality gives (a), (b) uses Jensen's inequality in view of the convexity of $f(x_1, \ldots, x_N) = (\prod_{i=1}^{N} x_i)^{-\xi}$ for $\xi > 0$, and the arithmetic-geometric mean inequality results in (c). Finally, the equivalence between maximizing the upper bound and maximizing $\sum_{l=1}^{n_r} \sum_{k=1}^{n_t} [\boldsymbol{\Omega}]_{i_l, j_k}$ trivially follows from the monotonicity of $\log_2(\cdot)$.

Appendix 4: Proof of Theorem 3.3

To facilitate the proof, we recall the following lemma.

Lemma 3.1 ([13]) *Let* $\mathbf{A} \in \mathbb{R}^{n\times n}$ *be symmetric and suppose that* $1 \leq m \leq n$. *Then*

$$\lambda_{\mathbf{A},[n-m+1]} + \cdots + \lambda_{\mathbf{A},[n]} = \min_{\substack{\mathbf{X}\in\mathbb{R}^{n\times m}\\ \mathbf{X}^{\mathrm{T}}\mathbf{X}=\mathbf{I}_m}} \operatorname{tr}\left(\mathbf{X}^{\mathrm{T}}\mathbf{A}\mathbf{X}\right),$$

where the minimum is achieved for a matrix $\mathbf{X}$ *whose columns are orthonormal eigenvectors associated with the* m *smallest eigenvalues of* $\mathbf{A}$.

We first establish the relation between $\operatorname{vec}(\cdot)$ and $\widetilde{\operatorname{vec}}(\cdot)$ in matrix terms. By introducing binary matrices $\overline{\mathbf{L}} = [\mathbf{L}_{\mathrm{d}}, \mathbf{L}_{\mathrm{l}}, \mathbf{L}_{\mathrm{u}}] \in \mathbb{R}^{N^2\times N^2}$, we may rewrite the $\operatorname{vec}(\cdot)$ operator for a Hermitian matrix $\mathbf{A} \in \mathbb{C}^{N\times N}$ as

$$\operatorname{vec}(\mathbf{A}) = [\mathbf{L}_{\mathrm{d}}, \mathbf{L}_{\mathrm{l}}, \mathbf{L}_{\mathrm{u}}]\left[\operatorname{diag}(\mathbf{A})^{\mathrm{T}}, \operatorname{vec}_{\mathrm{l}}^{\mathrm{T}}(\mathbf{A}), \operatorname{vec}_{\mathrm{l}}^{\mathrm{T}}\left(\mathbf{A}^*\right)\right]^{\mathrm{T}}$$

where $\overline{\mathbf{L}}$ satisfies $\overline{\mathbf{L}}\overline{\mathbf{L}}^{\mathrm{T}} = \mathbf{I}_{N^2}$ [27]. On the other hand, define a unitary matrix $\overline{\mathbf{T}} = \operatorname{diag}\{\mathbf{I}_N, \mathbf{T}\}$ with $\mathbf{T} = \frac{1}{\sqrt{2}}\begin{bmatrix} \mathbf{I}_{\frac{N(N-1)}{2}} & \jmath\mathbf{I}_{\frac{N(N-1)}{2}} \\ \mathbf{I}_{\frac{N(N-1)}{2}} & -\jmath\mathbf{I}_{\frac{N(N-1)}{2}} \end{bmatrix}$. It is straightforward to verify that

$$\widetilde{\operatorname{vec}}(\mathbf{A}) = \overline{\mathbf{T}}^{\mathrm{H}}\left[\operatorname{diag}(\mathbf{A})^{\mathrm{T}}, \operatorname{vec}_{\mathrm{l}}^{\mathrm{T}}(\mathbf{A}), \operatorname{vec}_{\mathrm{l}}^{\mathrm{T}}\left(\mathbf{A}^*\right)\right]^{\mathrm{T}}.$$

Thus, $\widetilde{\operatorname{vec}}(\mathbf{A}) = \overline{\mathbf{T}}^{\mathrm{H}}\overline{\mathbf{L}}^{\mathrm{T}}\operatorname{vec}(\mathbf{A})$. Now we show that $\mathbf{Y}^{\mathrm{T}}\mathbf{Y} = \mathbf{I}_{n_{\mathrm{t}}}$. Using $\|\mathbf{b}_k\| = 1$, it can be verified that $\widetilde{\operatorname{vec}}\left(\mathbf{b}_k\mathbf{b}_k^{\mathrm{H}}\right)$ has unit norm, i.e.,

$$\left\|\widetilde{\operatorname{vec}}\left(\mathbf{b}_k\mathbf{b}_k^{\mathrm{H}}\right)\right\|^2 = \left\|\operatorname{vec}\left(\mathbf{b}_k\mathbf{b}_k^{\mathrm{H}}\right)\right\|^2 = \left\|\mathbf{b}_k\mathbf{b}_k^{\mathrm{H}}\right\|_{\mathrm{F}}^2 = \operatorname{tr}\left(\mathbf{b}_k\mathbf{b}_k^{\mathrm{H}}\mathbf{b}_k\mathbf{b}_k^{\mathrm{H}}\right) = 1.$$

Further, from the identity $\mathrm{tr}\left(\mathbf{A}^{\mathrm{T}}\mathbf{B}\right) = \mathrm{vec}^{\mathrm{T}}(\mathbf{A})\,\mathrm{vec}(\mathbf{B})$ and $\mathbf{b}_i^{\mathrm{H}}\mathbf{b}_k = \delta_{ik}$, it follows that

$$\begin{aligned}\widetilde{\mathrm{vec}}^{\mathrm{T}}\left(\mathbf{b}_k\mathbf{b}_k^{\mathrm{H}}\right)\widetilde{\mathrm{vec}}\left(\mathbf{b}_l\mathbf{b}_l^{\mathrm{H}}\right) &= \mathrm{vec}^{\mathrm{H}}\left(\mathbf{b}_k\mathbf{b}_k^{\mathrm{H}}\right)\mathrm{vec}\left(\mathbf{b}_l\mathbf{b}_l^{\mathrm{H}}\right)\\ &= \mathrm{tr}\left[\left(\mathbf{b}_k^*\mathbf{b}_k^{\mathrm{T}}\right)^{\mathrm{T}}\mathbf{b}_l\mathbf{b}_l^{\mathrm{H}}\right]\\ &= \mathrm{tr}\left(\mathbf{b}_k\mathbf{b}_k^{\mathrm{H}}\mathbf{b}_l\mathbf{b}_l^{\mathrm{H}}\right)\\ &= 0.\end{aligned}$$

In conclusion, $\{\widetilde{\mathrm{vec}}(\mathbf{b}_k\mathbf{b}_k^{\mathrm{H}})\}$ is orthogonal, and a choice of semi-orthogonal $\mathbf{Y}$ that minimizes $\mathcal{J}_{\mathrm{Tx}}^{\mathrm{ub}}$ simply derives from Lemma 3.1. In addition, note that the objective function $\mathcal{J}_{\mathrm{Tx}}^{\mathrm{ub}} = \mathrm{tr}(\mathbf{Y}^{\mathrm{T}}\widetilde{\mathbf{G}}\widetilde{\mathbf{G}}^{\mathrm{T}}\mathbf{Y})$ is invariant under unitary rotation, i.e., for an arbitrary orthogonal matrix $\mathbf{Q} \in \mathbb{R}^{n_{\mathrm{t}}\times n_{\mathrm{t}}}$, it holds

$$\mathrm{tr}\left(\mathbf{Y}^{\mathrm{T}}\widetilde{\mathbf{G}}\widetilde{\mathbf{G}}^{\mathrm{T}}\mathbf{Y}\right) = \mathrm{tr}\left((\mathbf{Y}\mathbf{Q})^{\mathrm{T}}\widetilde{\mathbf{G}}\widetilde{\mathbf{G}}^{\mathrm{T}}\mathbf{Y}\mathbf{Q}\right).$$

Therefore, the optimal $\mathbf{Y}_{\mathrm{opt}}$ is a subspace spanned by the eigenvectors of $\widetilde{\mathbf{G}}\widetilde{\mathbf{G}}^{\mathrm{T}}$ associated with the n_{t} smallest eigenvalues.

References

1. O. El Ayach, S. Rajagopal, S. Abu-Surra, Z. Pi, and R. W. Heath, “Spatially sparse precoding in millimeter wave MIMO systems,” *IEEE Trans. Wireless Commun.*, vol. 13, no. 3, pp. 1499–1513, Mar. 2014.
2. X. Yu, J.-C. Shen, J. Zhang, and K. B. Letaief, “Alternating minimization algorithms for hybrid precoding in millimeter wave MIMO systems,” *IEEE J. Sel. Topics Signal Process.*, vol. 10, no. 3, pp. 485–500, Apr. 2016.
3. W. Weichselberger, M. Herdin, H. Özcelik, and E. Bonek, “A stochastic MIMO channel model with joint correlation of both link ends,” *IEEE Trans. Wireless Commun.*, vol. 5, no. 1, pp. 90–100, Jan. 2006.
4. A. M. Tulino, A. Lozano, and S. Verdú, “Capacity-achieving input covariance for single-user multi-antenna channels,” *IEEE Trans. Wireless Commun.*, vol. 5, no. 3, pp. 662–671, Mar. 2006.
5. X. Q. Gao, B. Jiang, X. Li, A. B. Gershman, and M. R. McKay, “Statistical eigenmode transmission over jointly correlated MIMO channels,” *IEEE Trans. Inf. Theory*, vol. 55, no. 8, pp. 3735–3750, Aug. 2009.
6. E. A. Jorswieck, R. Mochaourab, and M. Mittelbach, “Effective capacity maximization in multi-antenna channels with covariance feedback,” *IEEE Trans. Wireless Commun.*, vol. 9, no. 10, pp. 2988–2993, Oct. 2010.
7. J.-F. Cardoso and A. Souloumiac, “Blind beamforming for non-Gaussian signals,” *IEE Proc. Radar and Signal Process. F*, vol. 140, no. 6, pp. 362–370, Dec. 1993.
8. J. Sheinvald, “On blind beamforming for multiple non-Gaussian signals and the constant-modulus algorithm,” *IEEE Trans. Signal Process.*, vol. 46, no. 7, pp. 1878–1885, Jul. 1998.

9. E. Dahlman, S. Parkvall, and J. Sköld, *5G NR: The Next Generation Wireless Access Technology*. Cambridge, MA, USA: Academic Press, 2018.
10. A. M. Sayeed, "Deconstructing multiantenna fading channels," *IEEE Trans. Signal Process.*, vol. 50, no. 10, pp. 2563–2579, Oct. 2002.
11. D.-S. Shiu, G. J. Foschini, M. J. Gans, and J. M. Kahn, "Fading correlation and its effect on the capacity of multielement antenna systems," *IEEE Trans. Commun.*, vol. 48, no. 3, pp. 502–513, Mar. 2000.
12. D. P. Palomar, J. M. Cioffi, and M. A. Lagunas, "Joint Tx-Rx beamforming design for multicarrier MIMO channels: A unified framework for convex optimization," *IEEE Trans. Signal Process.*, vol. 51, no. 9, pp. 2381–2401, Sep. 2003.
13. R. A. Horn and C. R. Johnson, *Matrix Analysis*, 2nd ed. New York, NY, USA: Cambridge University Press, 2013.
14. D. Wu and R. Negi, "Effective capacity: A wireless link model for support of quality of service," *IEEE Trans. Wireless Commun.*, vol. 2, no. 4, pp. 630–643, Jul. 2003.
15. ——, "Effective capacity-based quality of service measures for wireless networks," *Mobile Netw. and Appl.*, vol. 11, no. 1, pp. 91–99, Feb. 2006.
16. M. C. Gursoy, "MIMO wireless communications under statistical queueing constraints," *IEEE Trans. Inf. Theory*, vol. 57, no. 9, pp. 5897–5917, Sep. 2011.
17. L. Vandenberghe and S. Boyd, "Semidefinite programming," *SIAM Review*, vol. 38, no. 1, pp. 49–95, Mar. 1996.
18. CVX Research, Inc., "CVX: Matlab software for disciplined convex programming, version 2.0," http://cvxr.com/cvx, Aug. 2012.
19. S. Joshi and S. Boyd, "Sensor selection via convex optimization," *IEEE Trans. Signal Process.*, vol. 57, no. 2, pp. 451–462, Feb. 2009.
20. J.-F. Cardoso and A. Souloumiac, "Jacobi angles for simultaneous diagonalization," *SIAM J. Mat. Anal. Appl.*, vol. 17, no. 1, pp. 161–164, Jan. 1996.
21. Q. U. Nadeem, A. Kammoun, M. Debbah, and M. S. Alouini, "A generalized spatial correlation model for 3D MIMO channels based on the Fourier coefficients of power spectrums," *IEEE Trans. Signal Process.*, vol. 63, no. 14, pp. 3671–3686, Jul. 2015.
22. ——, "3D massive MIMO systems: Modeling and performance analysis," *IEEE Trans. Wireless Commun.*, vol. 14, no. 2, pp. 6926–6939, Dec. 2015.
23. M. R. Akdeniz, Y. Liu, M. K. Samimi, S. Sun, S. Rangan, T. S. Rappaport, and E. Erkip, "Millimeter wave channel modeling and cellular capacity evaluation," *IEEE J. Sel. Areas Commun.*, vol. 32, no. 6, pp. 1164–1179, Jun. 2014.
24. X. Zhang, A. F. Molisch, and S.-Y. Kung, "Variable-phase-shift-based RF-baseband codesign for MIMO antenna selection," *IEEE Trans. Signal Process.*, vol. 53, no. 11, pp. 4091–4103, Nov. 2005.
25. J. T. Chen and V. K. N. Lau, "Two-tier precoding for FDD multi-cell massive MIMO time-varying interference networks," *IEEE J. Sel. Areas Commun.*, vol. 32, no. 6, pp. 1230–1238, Jun. 2014.
26. D. Hösli and A. Lapidoth, "The capacity of a MIMO Ricean channel is monotonic in the singular values of the mean," in *Proc. 5th Intl. ITG Conf. Source and Channel Coding*, Erlangen, Germany, Jan. 2004.
27. A. Hjørungnes, *Complex-Valued Matrix Derivatives With Applications in Signal Processing and Communications*. New York, NY, USA: Cambridge University Press, 2011.

Chapter 4
Hybrid Precoding/Combining for Massive MIMO with Hybrid ARQ

4.1 Introduction

Because of poor channel conditions, correct reception at a certain target data rate can sometimes become impossible. Therefore, packet retransmission protocols are used in modern wireless communications systems to improve the reliability of data transmission. Specifically, when a one-bit feedback link is available from the receiver to the transmitter, the simple scheme of hybrid automatic repeat request (ARQ) with Chase combining (HARQ-CC) can be employed, where the same packet is retransmitted in the event of decoding failure.

In this chapter, we would like to address the design of hybrid RF-baseband precoding and combining suitable for massive multiple-input multiple-output (MIMO) with hybrid ARQ. In particular, based on the perfect channel state information (CSI), we consider a progressive approach to the hybrid precoder and combiner design in an attempt to exploit the temporal diversity inherent in packet retransmissions. Conditioned on the knowledge of previous retransmissions, the hybrid precoder and combiner are sequentially optimized for the current ARQ round without considering potential future retransmissions. To this end, we propose a two-step strategy to optimize the hybrid RF-baseband precoders/combiners with the objective of maximizing the spectral efficiency. On the heuristic assumption that the linear minimum mean square error (MMSE) filter is perfectly realizable by the two-stage receive combiner, we separate the derivation of the hybrid precoding from the hybrid combining. At the transmitter, during each ARQ round, we choose the RF precoder either from the set of transmit array response vectors or from a discrete Fourier transform (DFT)-based codebook. Built upon the selected RF precoder, the optimal beamformer at baseband is analytically shown to be a function of the generalized eigenvectors of the effective channel and RF precoder for the current ARQ retransmission, while the transmit power is allocated based on the precoding solutions from the previous and current packet retransmissions. At the receiver, a novel hybrid combining structure is proposed to address the issue

T. Le-Ngoc, R. Mai, *Hybrid Massive MIMO Precoding in Cloud-RAN*,
Wireless Networks, https://doi.org/10.1007/978-3-030-02158-0_4

of increased computational and storage complexity caused by repeated packet retransmissions. To minimize the performance loss of the decoupled precoding-combining optimization, the two-step strategy is further applied to derive the hybrid combining solution as an approximation of the optimal linear digital counterpart in terms of error performance. Through numerical simulations, we validate the efficacy of the proposed progressive approach to hybrid precoding and combining from (1) its comparable performance with the fully digital optimal progressive scheme, and (2) its performance advantage over other hybrid baseline that is oblivious to the presence of time diversity.

Although previous works, e.g., [1–3], have proposed various solution techniques for hybrid precoding and combining in the context of massive MIMO, they do not directly carry over when hybrid ARQ is incorporated. The study in this chapter is motivated by the observation that sequential precoding optimization is unique to ARQ systems as data retransmission occurs only if the previous transmissions fail. Since it is not possible to alter the previous transmission attempts of a data packet and another retransmission attempt may not be required after the current transmission, the transmitter can only optimize the current transmission for each ARQ round with respect to the desired performance metric. In theory, the formulation of matrix reconstruction as in Chap. 3 can be exploited to solve for the hybrid design as an approximation of the optimal solution. This nonetheless requires the knowledge of the fully digital solutions, which is generally nontrivial to obtain in the first place. Furthermore, such an approach tends to lead to sub-optimal power loading schemes at baseband. In contrast, the proposed progressive precoding solution takes into account power allocation for each round of packet retransmission, and also gives insights into the relation between the precoder for the current ARQ transmission and the previous ones. Mathematically, the presence of the RF stage in the power constraint also necessitates a different solution approach.

The rest of this chapter is organized as follows. In Sect. 4.2, we describe the system model of massive MIMO with hybrid ARQ where hybrid precoding and combining are employed. We develop the proposed methods of progressive optimization of hybrid precoding and combining in Sects. 4.3 and 4.4, respectively. Illustrative results are provided in Sect. 4.5, where the proposed progressive hybrid solution is numerically compared with various baselines. Concluding remarks are made in Sect. 4.6.

4.2 System Model and Problem Statement

We consider a point-to-point flat-fading massive MIMO system with N_t antennas at the transmitter and N_r antennas at the receiver, where the large-scale transmit and receive antenna arrays are driven by $n_\mathrm{t} \ll N_\mathrm{t}$ and $n_\mathrm{r} \ll N_\mathrm{r}$ RF chains, respectively. Let $N_\mathrm{s} \leq \min(n_\mathrm{t}, n_\mathrm{r})$ denote the number of data streams to be transmitted. As a result of a limited number of RF chains, a hybrid RF-baseband precoder/combiner

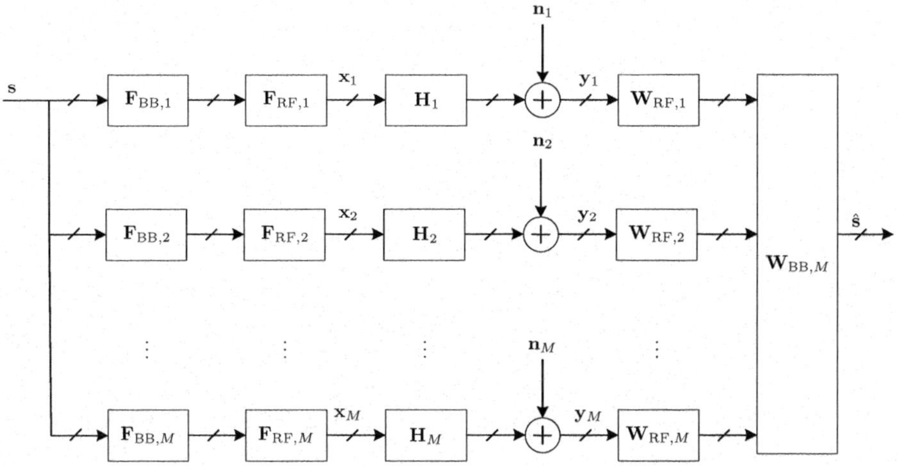

Fig. 4.1 Hybrid RF-baseband precoding and combining for M rounds of packet retransmission in a point-to-point massive MIMO hybrid ARQ system

in place of a single-stage, fully digital precoder/combiner is employed. As illustrated in Fig. 4.1, the data streams are first linearly transformed using a low-dimensional baseband precoder $\mathbf{F}_{\mathrm{BB}} \in \mathbb{C}^{n_{\mathrm{t}} \times N_{\mathrm{s}}}$, the output of which is then precoded by a high-dimensional RF precoder $\mathbf{F}_{\mathrm{RF}} \in \mathbb{C}^{N_{\mathrm{t}} \times n_{\mathrm{t}}}$. The RF precoder is assumed to be implemented using analog phase-shifters, i.e., the elements are constrained to satisfy $|[\mathbf{F}_{\mathrm{RF}}]_{i,j}| = \frac{1}{\sqrt{N_{\mathrm{t}}}}$, $1 \le i \le N_{\mathrm{t}}$, $1 \le j \le n_{\mathrm{t}}$.

At the receiver, a single-bit feedback is leveraged to inform the transmitter of the reception status. In particular, the receiver sends an acknowledgment (ACK) for successful decoding and a negative ACK (NACK) signal otherwise. For the case of HARQ-CC which is of interest to this work, the transmitter simply resends the same signal upon receiving a NACK feedback. This makes it possible for the receiver to combine the same transmitted signal across all transmission attempts, as shown in Fig. 4.1. Denoting the RF and baseband precoders during the mth ARQ round of transmission by $\mathbf{F}_{\mathrm{RF},m}$ and $\mathbf{F}_{\mathrm{BB},m}$, respectively, the received signal during the mth ARQ transmission is given by

$$\mathbf{y}_m = \sqrt{\rho}\mathbf{H}_m\mathbf{F}_{\mathrm{RF},m}\mathbf{F}_{\mathrm{BB},m}\mathbf{s} + \mathbf{n}_m,$$

where ρ represents the average received power, the data vector $\mathbf{s}$ is assumed to be encoded by an independently and identically distributed (i.i.d.) Gaussian codebook, i.e., $\mathbf{s} \sim \mathcal{CN}(\mathbf{0}, \frac{1}{N_{\mathrm{s}}}\mathbf{I}_{N_{\mathrm{s}}})$, $\mathbf{H}_m \in \mathbb{C}^{N_{\mathrm{r}} \times N_{\mathrm{t}}}$ denotes the massive MIMO channel between the transmitter and receiver during the mth retransmission of the data vector $\mathbf{s}$, and is assumed to change independently from transmission to transmission, and $\mathbf{n}_m \sim \mathcal{CN}\left(\mathbf{0}, \sigma_n^2\mathbf{I}_{N_{\mathrm{r}}}\right)$ is the spatially white Gaussian noise with variance σ_n^2. In particular, we characterize $\mathbf{H}_m, \forall m$ by a parametric clustered channel model [1, 4], in which the channel matrix is a sum of contributions of N_{cl} scattering clusters, each with N_{ray} propagation paths. Assuming uniform linear arrays (ULA) at both link ends, the MIMO channel is generically expressed as[1]

$$\mathbf{H} = \gamma \sum_{i=1}^{N_{\text{cl}}} \sum_{l=1}^{N_{\text{ray}}} \alpha_{il} \mathbf{a}_{\text{r}} \left(\phi_{il}^{\text{r}}\right) \mathbf{a}_{\text{t}} \left(\phi_{il}^{\text{t}}\right)^{\text{H}},$$

where $\gamma = \sqrt{N_{\text{t}} N_{\text{r}} / N_{\text{cl}} N_{\text{ray}}}$ is a normalization factor such that $\mathbb{E}[\|\mathbf{H}\|_{\text{F}}^2] = N_{\text{r}} N_{\text{t}}$, and $\alpha_{il} \sim \mathcal{CN}(0, 1)$ is the i.i.d. complex path gain of the lth ray in the ith scattering cluster. The vector $\mathbf{a}_{\text{r}} \left(\phi_{il}^{\text{r}}\right) \in \mathbb{C}^{N_{\text{r}}}$ ($\mathbf{a}_{\text{t}} \left(\phi_{il}^{\text{t}}\right) \in \mathbb{C}^{N_{\text{t}}}$) represents the normalized receive (transmit) array response vector at an azimuth angle of arrival (AoA) ϕ_{il}^{r} (angle of departure (AoD) ϕ_{il}^{t}). For a generic N-element ULA on the y-axis, the array response is expressed as [1]

$$\mathbf{a}_{\text{ULA}} (\phi) = \frac{1}{\sqrt{N}} \left[1, e^{\jmath 2\pi d_\lambda \sin(\phi)}, \dots, e^{\jmath (N-1) 2\pi d_\lambda \sin(\phi)}\right]^{\text{T}}, \tag{4.1}$$

where d_λ is the inter-antenna spacing normalized by the wavelength. The perfect knowledge of $\mathbf{H}_m, \forall m$ is assumed to be available at the transmitter and receiver. While it is challenging to acquire the high-dimensional CSI in closed-loop frequency-division duplexing (FDD) systems which depend on the mechanism of channel training and feedback, such an issue can be simplified by exploiting channel reciprocity in time-division duplexing (TDD) systems.[2]

After M retransmissions of the data vector $\mathbf{s}$, the received signals can be aggregately written as

[1]In this work, we consider hybrid precoding and combining in the azimuth plane, and therefore restrict our attention to the use of ULAs at the transmitter and receiver. Nonetheless, we note that the proposed design techniques directly carry over to the case of full-dimensional beamforming where both azimuth and elevation beamforming are enabled by arranging the antenna elements in a uniform rectangular configuration or a cylindrical configuration.

[2]For the channel reciprocity to be useful, it is required that the sample duration allocated for uplink channel training T_u and that for downlink data transmission T_d should not exceed the channel coherence time T_{coh}. Suppose that the channel remains invariant before the user moves a quarter of the wavelength λ. Accordingly, the channel coherence time is calculated as $T_{\text{coh}} = \frac{1}{4}\frac{\lambda}{v}$ with v as the mobile speed of the terminal. Furthermore, let T_s denote the sample duration of an orthogonal frequency-division multiplexing (OFDM) symbol, which consists of a guard interval T_{cp} and an effective symbol interval T_{eff}. To avoid inter-symbol interference, it can be assumed for simplicity that the guard interval is set equal to the channel delay spread τ, i.e., $T_{\text{cp}} = \tau$. Accordingly, the channel coherence time in terms of the number of channel uses is expressed by $T_{\text{coh}}^{\text{cu}} = \frac{T_{\text{coh}} T_{\text{eff}}}{T_s T_{\text{cp}}}$. To estimate the frequency response, the N_{r} transceivers at the user terminal transmit orthogonal pilot sequences on the uplink, each of length no less than N_{r} channel uses. In general, it is required that $T_{\text{coh}}^{\text{cu}} \geq 2N_{\text{r}}$ such that the ensuing data transmission could take place [5].

$$\tilde{\mathbf{y}}_M = \begin{bmatrix} \mathbf{y}_1 \\ \mathbf{y}_2 \\ \vdots \\ \mathbf{y}_M \end{bmatrix} = \sqrt{\rho} \underbrace{\begin{bmatrix} \mathbf{H}_1\mathbf{F}_{\mathrm{RF},1}\mathbf{F}_{\mathrm{BB},1} \\ \mathbf{H}_2\mathbf{F}_{\mathrm{RF},2}\mathbf{F}_{\mathrm{BB},2} \\ \vdots \\ \mathbf{H}_M\mathbf{F}_{\mathrm{RF},M}\mathbf{F}_{\mathrm{BB},M} \end{bmatrix}}_{\triangleq \mathbb{H}} \mathbf{s} + \underbrace{\begin{bmatrix} \mathbf{n}_1 \\ \mathbf{n}_2 \\ \vdots \\ \mathbf{n}_M \end{bmatrix}}_{\triangleq \tilde{\mathbf{n}}_M}$$

$$= \sqrt{\rho}\mathbb{H}\mathbf{s} + \tilde{\mathbf{n}}_M.$$

As the noise vectors during different retransmissions are independent, it holds that $\mathbf{R}_{\tilde{n}} \triangleq \mathbb{E}[\tilde{\mathbf{n}}_M\tilde{\mathbf{n}}_M^{\mathrm{H}}] = \sigma_n^2\mathbf{I}_{MN_{\mathrm{r}}}$. The fact that the same signal is transmitted in the event of decoding failure enables the coherent combining of all the received signals across different ARQ rounds. Ideally, the combining should take place in both the RF and baseband domains for the optimal performance, which would require storage of the high-dimensional received signals $\{\mathbf{y}_m\}_{m=1}^M$ from all ARQ rounds and applying to the aggregate received signal $\tilde{\mathbf{y}}_M$ an RF combiner of size $MN_{\mathrm{r}} \times Mn_{\mathrm{r}}$ followed by a baseband combiner of size $Mn_{\mathrm{r}} \times N_{\mathrm{s}}$. Unfortunately, these high-dimensional combiners pose nontrivial computational and storage complexity, especially when N_{r} is large. Furthermore, additional RF phase-shifters are needed to implement the joint combining in the RF domain.[3] Out of these concerns, in this work, we propose to first perform independent RF combining of the received signals from each ARQ round, and then perform joint baseband combining of the RF-processed received signals across all ARQ rounds, as illustrated in Fig. 4.1. Accordingly, after the Mth ARQ round, the signal at the output of the hybrid combiner can be written as

$$\hat{\mathbf{y}}_M = \mathbf{W}_{\mathrm{BB},M}^{\mathrm{H}}\mathbb{W}_{\mathrm{RF},M}^{\mathrm{H}}\tilde{\mathbf{y}}_M = \sqrt{\rho}\mathbf{W}_{\mathrm{BB},M}^{\mathrm{H}}\mathbb{W}_{\mathrm{RF},M}^{\mathrm{H}}\mathbb{H}\mathbf{s} + \mathbf{W}_{\mathrm{BB},M}^{\mathrm{H}}\mathbb{W}_{\mathrm{RF},M}^{\mathrm{H}}\tilde{\mathbf{n}}_M,$$

where the block diagonal matrix $\mathbb{W}_{\mathrm{RF},M} = \mathrm{blkdiag}\{\mathbf{W}_{\mathrm{RF},1}, \ldots, \mathbf{W}_{\mathrm{RF},M}\}$ is the aggregate RF combiner with the mth component $\mathbf{W}_{\mathrm{RF},m} \in \mathbb{C}^{N_{\mathrm{r}} \times n_{\mathrm{r}}}$, $1 \le m \le M$, as the RF combiner with respect to the received signal $\mathbf{y}_m$, and $\mathbf{W}_{\mathrm{BB},M} \in \mathbb{C}^{Mn_{\mathrm{r}} \times N_{\mathrm{s}}}$ is a baseband combiner used during the Mth ARQ round. The achievable rate per channel use, i.e., spectral efficiency, with the proposed precoding and combining strategy after M ARQ rounds is therefore expressed as

$$R_{ARQ}^{M} = \frac{1}{M}\log_2\left|\mathbf{I}_{N_{\mathrm{s}}} + \beta\mathbf{R}_{\mathbf{W}_n}^{-1}\mathbf{W}_{\mathrm{BB},M}^{\mathrm{H}}\mathbb{W}_{\mathrm{RF},M}^{\mathrm{H}}\mathbb{H}\mathbb{H}^{\mathrm{H}}\mathbb{W}_{\mathrm{RF},M}\mathbf{W}_{\mathrm{BB},M}\right|, \tag{4.2}$$

where we define $\beta \triangleq \frac{\rho}{N_{\mathrm{s}}\sigma_n^2}$ and $\mathbf{R}_{\mathbf{W}_n} \triangleq \mathbf{W}_{\mathrm{BB},M}^{\mathrm{H}}\mathbb{W}_{\mathrm{RF},M}^{\mathrm{H}}\mathbb{W}_{\mathrm{RF},M}\mathbf{W}_{\mathrm{BB},M}$ for notational simplicity.

[3] In an ARQ-based system with a single-channel (and a single processing unit), multiple copies of the same signal are received independently at different time instances. However, in a multi-channel system, performing combining across the independently received signals from parallel channels can take place as each channel will have a separate processing unit.

4.3 Progressive Hybrid Precoding Design

In this section, we formulate the progressive hybrid precoding problem during the Mth ARQ round of data retransmission. Built upon the hybrid precoders $\{\mathbf{F}^{\star}_{\mathrm{RF},m}, \mathbf{F}^{\star}_{\mathrm{BB},m}\}_{m=1}^{M-1}$ and hybrid combiners $\{\mathbf{W}^{\star}_{\mathrm{RF},m}, \mathbf{W}^{\star}_{\mathrm{BB},m}\}_{m=1}^{M-1}$ used for the previous retransmission attempts, the objective is to find the hybrid precoder $\{\mathbf{F}_{\mathrm{RF},M}, \mathbf{F}_{\mathrm{BB},M}\}$ and hybrid combiner $\{\mathbf{W}_{\mathrm{RF},M}, \mathbf{W}_{\mathrm{BB},M}\}$ during the Mth ARQ round such that R^{M}_{ARQ} in (4.2) is maximized.

Finding the global optimum generally requires joint hybrid precoding and combining optimization, which is unfortunately mathematically intractable especially in the presence of the nonconvex modulus constraints on $\mathbf{F}_{\mathrm{RF},M}$ and $\mathbf{W}_{\mathrm{RF},M}$. Instead, we seek an approach where the hybrid precoding can be decoupled from the hybrid combining optimization. In particular, we assume that the fully digital linear MMSE combiner is perfectly realizable by the hybrid RF-baseband counterpart. Following a similar line of reasoning as in [6], it can be shown that in this case, the achievable rate R^{M}_{ARQ} in (4.2) is reduced to the mutual information between $\mathbf{s}$ and $\tilde{\mathbf{y}}_M$, i.e., $\mathcal{I}(\mathbf{s}; \tilde{\mathbf{y}}_M)$, as given by

$$\mathcal{I}\left(\mathbf{s}; \tilde{\mathbf{y}}_M\right) = \log_2 \Bigg| \mathbf{I}_{N_s} + \beta \Bigg(\sum_{m=1}^{M-1} \mathbf{F}^{\star\mathrm{H}}_{\mathrm{BB},m} \mathbf{F}^{\star\mathrm{H}}_{\mathrm{RF},m} \mathbf{H}^{\mathrm{H}}_{m} \mathbf{H}_m \mathbf{F}^{\star}_{\mathrm{RF},m} \mathbf{F}^{\star}_{\mathrm{BB},m} + \mathbf{F}^{\mathrm{H}}_{\mathrm{BB},M} \mathbf{F}^{\mathrm{H}}_{\mathrm{RF},M} \mathbf{H}^{\mathrm{H}}_{M} \mathbf{H}_M \mathbf{F}_{\mathrm{RF},M} \mathbf{F}_{\mathrm{BB},M} \Bigg) \Bigg|,$$

where we note that optimization of the precoders $\mathbf{F}_{\mathrm{BB},M}$ and $\mathbf{F}_{\mathrm{RF},M}$ for the Mth ARQ round depends on the previous precoders $\{\mathbf{F}^{\star}_{\mathrm{RF},m}, \mathbf{F}^{\star}_{\mathrm{BB},m}\}_{m=1}^{M-1}$. The problem of interest is accordingly formulated as

$$\max_{\mathbf{F}_{\mathrm{RF},M}, \mathbf{F}_{\mathrm{BB},M}} \quad \mathcal{I}\left(\mathbf{s}; \tilde{\mathbf{y}}_M\right) \qquad \text{s.t.} \quad \mathbf{F}_{\mathrm{RF},M} \in \mathcal{F}_{\mathrm{RF}}, \; \left\| \mathbf{F}_{\mathrm{RF},M} \mathbf{F}_{\mathrm{BB},M} \right\|_{\mathrm{F}}^{2} = N_{\mathrm{S}}. \tag{4.3}$$

In view of the difficulty with jointly solving for $\{\mathbf{F}_{\mathrm{RF},M}, \mathbf{F}_{\mathrm{BB},M}\}$ subject to the constraint $\mathbf{F}_{\mathrm{RF},M} \in \mathcal{F}_{\mathrm{RF}}$, we propose a two-step solution technique. In the first step, we choose the RF precoder $\mathbf{F}_{\mathrm{RF},M}$ to be one of the feasible solutions from the set $\mathcal{F}_{\mathrm{RF}}$, say $\mathbf{G}_{\mathrm{RF},M}$, and conditioned on this choice, we derive the closed-form solution for the optimal baseband precoder $\mathbf{F}^{\star}_{\mathrm{BB},M}\left(\mathbf{G}_{\mathrm{RF},M}\right)$. This procedure is repeated for each feasible $\mathbf{G}_{\mathrm{RF},M}$ to generate a set of $\{\mathbf{G}_{\mathrm{RF},M}, \mathbf{F}^{\star}_{\mathrm{BB},M}(\mathbf{G}_{\mathrm{RF},M})\}$, from which the pair that yields the maximum mutual information in (4.3) is declared as the solution to the original problem. We remark that unlike the one-shot approach based on matrix reconstruction in [1], the proposed two-step technique obviates the need for knowing the fully digital solution, and enjoys the flexibility of performing waterfilling-based power loading at baseband.

Clearly, the set of constant-modulus RF precoders $\mathcal{F}_{\mathrm{RF}}$ contains an unlimited number of elements. To facilitate the RF precoding optimization, we consider two suboptimal but effective alternatives, where the columns of the RF precoder $\mathbf{F}_{\mathrm{RF},M}$ are assumed to be chosen either from the set of $N_{\mathrm{cl}}N_{\mathrm{ray}}$ transmit array response vectors $\mathcal{A}_{\mathrm{t}} \triangleq \{\mathbf{a}_{\mathrm{t}}(\phi_{il}^{\mathrm{t}})\}_{i,l}$ [1] or from the N_{t} columns of the N_{t}-dimensional DFT matrix $\mathcal{D}_{\mathrm{t}} \triangleq \left[\frac{1}{\sqrt{N_{\mathrm{t}}}}e^{-\frac{\jmath 2\pi mn}{N_{\mathrm{t}}}}\right], 0 \le m, n \le N_{\mathrm{t}} - 1$ [2]. The rationale for constraining the columns of the RF precoder to the finite sets $\mathcal{A}_{\mathrm{t}}$ and $\mathcal{D}_{\mathrm{t}}$ for the ARQ retransmission is threefold: (1) on one hand, the optimal progressive digital precoder has been known to be a function of a unitary matrix whose columns span the row space of the channel $\mathbf{H}_M$ [7]. On the other hand, it has been observed in [1] that under certain conditions, the set of transmit array response vectors $\mathcal{A}_{\mathrm{t}}$ serves as another basis for the row space of the channel $\mathbf{H}_M$; (2) as indicated by the definition (4.1), the transmit array response vector $\mathbf{a}_{\mathrm{t}}(\phi_{il}^{\mathrm{t}}), \forall i, l$, consists of constant-modulus entries; (3) when angle-domain quantization is considered as an option to reduce the overhead of estimating the complete AoD information, the array response vector-based RF precoding design can still be directly applied. One extreme case is the DFT-based codebook, which represents blind uniform quantization of the azimuth AoDs. Interestingly, in the large-scale array regime, the DFT matrix becomes an asymptotically good approximation for the channel eigen-space [8]. Our proposed approach to generate the hybrid RF-baseband precoder for the Mth ARQ retransmission is summarized in Algorithm 4.1.

Algorithm 4.1 Two-step progressive hybrid RF-baseband precoding design

1. Construct $\mathbf{G}_{\mathrm{RF},M}$ by choosing n_{t} vectors from either the set of $N_{\mathrm{cl}}N_{\mathrm{ray}}$ array response vectors $\mathcal{A}_{\mathrm{t}} = \{\mathbf{a}_{\mathrm{t}}(\phi_{il}^{\mathrm{t}})\}_{i,l}$ or the N_{t} columns of the DFT matrix $\mathcal{D}_{\mathrm{t}}$.
2. For the selected $\mathbf{G}_{\mathrm{RF},M}$, find the optimal $\mathbf{F}_{\mathrm{BB},M}^{\star}(\mathbf{G}_{\mathrm{RF},M})$ by solving the following optimization problem

$$\begin{aligned} \max_{\mathbf{F}_{\mathrm{BB},M}} \ & \log_2 \left| \mathbf{I}_{N_{\mathrm{s}}} + \beta \sum_{m=1}^{M-1} \mathbf{F}_{\mathrm{BB},m}^{\star \mathrm{H}} \mathbf{F}_{\mathrm{RF},m}^{\star \mathrm{H}} \mathbf{H}_m^{\mathrm{H}} \mathbf{H}_m \mathbf{F}_{\mathrm{RF},m}^{\star} \mathbf{F}_{\mathrm{BB},m}^{\star} \right. \\ & \left. \qquad + \beta \mathbf{F}_{\mathrm{BB},M}^{\mathrm{H}} \mathbf{G}_{\mathrm{RF},M}^{\mathrm{H}} \mathbf{H}_M^{\mathrm{H}} \mathbf{H}_M \mathbf{G}_{\mathrm{RF},M} \mathbf{F}_{\mathrm{BB},M} \right| \\ \text{s.t.} \ & \left\| \mathbf{G}_{\mathrm{RF},M} \mathbf{F}_{\mathrm{BB},M} \right\|_{\mathrm{F}}^2 = N_{\mathrm{s}} \end{aligned} \tag{4.4}$$

3. Repeat step 2 for each combination of $\mathbf{G}_{\mathrm{RF},M}$ from the set $\{\mathbf{a}_{\mathrm{t}}(\phi_{il}^{\mathrm{t}})\}_{i,l}$ or from the columns of $\mathcal{D}_t$ and choose the pair $\{\mathbf{G}_{\mathrm{RF},M}, \mathbf{F}_{\mathrm{BB},M}^{\star}(\mathbf{G}_{\mathrm{RF},M})\}$ that yields the maximum mutual information as the solution to (4.3).

In order to derive the optimal baseband solution $\mathbf{F}^{\star}_{\mathrm{BB},M}\left(\mathbf{G}_{\mathrm{RF},M}\right)$ to (4.4), we observe that given the eigenvalue decomposition

$$\sum_{m=1}^{M-1} \mathbf{F}^{\star \mathrm{H}}_{\mathrm{BB},m} \mathbf{F}^{\star \mathrm{H}}_{\mathrm{RF},m} \mathbf{H}^{\mathrm{H}}_{m} \mathbf{H}_{m} \mathbf{F}^{\star}_{\mathrm{RF},m} \mathbf{F}^{\star}_{\mathrm{BB},m} = \mathbf{U} \boldsymbol{\Lambda}_{M-1} \mathbf{U}^{\mathrm{H}},$$

the objective function in (4.4) can be reduced to

$$\begin{aligned} &\log_2 \left| \mathbf{I}_{N_\mathrm{s}} + \beta \mathbf{U} \boldsymbol{\Lambda}_{M-1} \mathbf{U}^{\mathrm{H}} + \beta \mathbf{F}^{\mathrm{H}}_{\mathrm{BB},M} \mathbf{G}^{\mathrm{H}}_{\mathrm{RF},M} \mathbf{H}^{\mathrm{H}}_{M} \mathbf{H}_{M} \mathbf{G}_{\mathrm{RF},M} \mathbf{F}_{\mathrm{BB},M} \right| \\ &= \log_2 \left| \widetilde{\boldsymbol{\Lambda}}_{M-1} + \beta \widetilde{\mathbf{F}}^{\mathrm{H}}_{\mathrm{BB},M} \mathbf{G}^{\mathrm{H}}_{\mathrm{RF},M} \mathbf{H}^{\mathrm{H}}_{M} \mathbf{H}_{M} \mathbf{G}_{\mathrm{RF},M} \widetilde{\mathbf{F}}_{\mathrm{BB},M} \right|, \end{aligned}$$

where $\widetilde{\boldsymbol{\Lambda}}_{M-1} \triangleq \mathbf{I}_{N_\mathrm{s}} + \beta \boldsymbol{\Lambda}_{M-1} = \mathrm{diag}\{\widetilde{\lambda}_{1,M-1}, \ldots, \widetilde{\lambda}_{N_\mathrm{s},M-1}\}$, and $\widetilde{\mathbf{F}}_{\mathrm{BB},M} \triangleq \mathbf{F}_{\mathrm{BB},M}\mathbf{U}$. On the other hand, following the orthonormality of $\mathbf{U}$, the power constraint in (4.4) is equivalent to $\|\mathbf{G}_{\mathrm{RF},M}\widetilde{\mathbf{F}}_{\mathrm{BB},M}\|_\mathrm{F}^2 = N_\mathrm{s}$. In other words, we can reformulate the problem (4.4) as

$$\begin{aligned} &\max_{\widetilde{\mathbf{F}}_{\mathrm{BB},M}} \log_2 \left| \widetilde{\boldsymbol{\Lambda}}_{M-1} + \beta \widetilde{\mathbf{F}}^{\mathrm{H}}_{\mathrm{BB},M} \mathbf{G}^{\mathrm{H}}_{\mathrm{RF},M} \mathbf{H}^{\mathrm{H}}_{M} \mathbf{H}_{M} \mathbf{G}_{\mathrm{RF},M} \widetilde{\mathbf{F}}_{\mathrm{BB},M} \right| \\ &\text{s.t.} \quad \left\| \mathbf{G}_{\mathrm{RF},M} \widetilde{\mathbf{F}}_{\mathrm{BB},M} \right\|_\mathrm{F}^2 = N_\mathrm{s}, \end{aligned} \tag{4.5}$$

to which the optimal solution can be analytically established, as shown in the following theorem.

Theorem 4.1 *Let* $\mathbf{A}_M$ *jointly diagonalize the pair of positive semi-definite matrices* $\mathbf{G}^{\mathrm{H}}_{\mathrm{RF},M}\mathbf{H}^{\mathrm{H}}_{M}\mathbf{H}_{M}\mathbf{G}_{\mathrm{RF},M}$ *and* $\mathbf{G}^{\mathrm{H}}_{\mathrm{RF},M}\mathbf{G}_{\mathrm{RF},M}$ *as*

$$\begin{aligned} \mathbf{A}^{\mathrm{H}}_{M} \mathbf{G}^{\mathrm{H}}_{\mathrm{RF},M} \mathbf{H}^{\mathrm{H}}_{M} \mathbf{H}_{M} \mathbf{G}_{\mathrm{RF},M} \mathbf{A}_{M} &= \boldsymbol{\Sigma}_{1M}, \\ \mathbf{A}^{\mathrm{H}}_{M} \mathbf{G}^{\mathrm{H}}_{\mathrm{RF},M} \mathbf{G}_{\mathrm{RF},M} \mathbf{A}_{M} &= \boldsymbol{\Sigma}_{2M}, \end{aligned}$$

where $\boldsymbol{\Sigma}_{iM} = \mathrm{diag}\{\sigma_{1,iM}, \ldots, \sigma_{n_\mathrm{t},iM}\}$, $i = 1, 2$. *Let* $k_1, \ldots, k_{N_\mathrm{s}}$ *denote the indices of the largest* N_s *elements of the set* $\{\psi_{i,M} \triangleq \frac{\sigma_{i,1M}}{\sigma_{i,2M}}\}_{i=1}^{n_\mathrm{t}}$ *such that* $\psi_{k_1,M} \geq \cdots \geq \psi_{k_{N_\mathrm{s}},M}$. *Denote* $\boldsymbol{\Psi}_M \triangleq \mathrm{diag}\{\psi_{k_1,M}, \ldots, \psi_{k_{N_\mathrm{s}},M}\}$. *Let* $\mathbf{P}_{1,M}$ *and* $\mathbf{P}_{2,M}$ *be the permutation matrices such that the diagonal elements of* $\mathbf{P}_{1,M}\boldsymbol{\Psi}_M\mathbf{P}^{\mathrm{H}}_{1,M}$ *are in non-decreasing order while those of* $\mathbf{P}_{2,M}\widetilde{\boldsymbol{\Lambda}}_{M-1}\mathbf{P}^{\mathrm{H}}_{2,M}$ *are in non-increasing order. Then the optimal solution to (4.5) is given by*

$$\widetilde{\mathbf{F}}_{\mathrm{BB},M} = \mathbf{A}_M \mathbf{S}_M \mathbf{P}^{\mathrm{H}}_{1,M} \widehat{\mathbf{F}}_{\mathrm{BB},M} \mathbf{P}_{2,M}$$

where $\widehat{\mathbf{F}}_{\mathrm{BB},M} = \mathrm{diag}\{f_M^1, \cdots, f_M^{N_\mathrm{s}}\}$ *with*

$$f_M^j = \sqrt{\frac{1}{\sigma_{k_j,2M}}\left(c - \frac{\widehat{\lambda}_{1,M-1}}{\beta\psi_{k_j,M}}\right)^+}, 1 \le j \le N_\mathrm{s},$$

and c chosen such that $\mathrm{tr}(\widehat{\mathbf{F}}_{\mathrm{BB},M}\widehat{\mathbf{F}}_{\mathrm{BB},M}^{\mathrm{H}}) = N_\mathrm{s}$*, and* $\mathbf{S}_M$ *selects the* N_s *columns of* $\mathbf{A}_M$ *indexed by* $k_1, \ldots, k_{N_\mathrm{s}}$*.*

Proof See Appendix 1.

Remark 4.1 The optimal baseband precoder, conditioned on the selected RF precoder $\mathbf{G}_{\mathrm{RF},M}$, consists of the following features:

1. The optimal beamforming subspace is spanned by the generalized eigenvectors $\mathbf{A}_M$ of the Gram matrices of the effective channel $\mathbf{H}_M\mathbf{G}_{\mathrm{RF},M}$ and the RF precoder $\mathbf{G}_{\mathrm{RF},M}$.
2. The optimal beamforming directions are determined by the selection matrix $\mathbf{S}_M$, which picks the N_s largest indices of the set $\{\psi_{i,M}\}_{i=1}^{n_\mathrm{t}}$.
3. The diagonal matrix $\widehat{\mathbf{F}}_{\mathrm{BB},M}$ implements the waterfilling-based power loading.
4. The permutation matrices $\mathbf{P}_{1,M}$ and $\mathbf{P}_{2,M}$ achieve the optimal reverse pairing of the singular values of the *previous* precoded transmissions $\mathbf{I}_{N_\mathrm{s}} + \beta\mathbf{\Lambda}_{M-1}$ with the largest N_s elements of the set $\{\psi_{1,M}, \ldots, \psi_{n_\mathrm{t},M}\}$ related to the *current* transmission.
5. For the initial transmission, i.e., $M = 1$, the permutation matrices become an identity matrix, i.e., $\mathbf{P}_{1,M} = \mathbf{P}_{2,M} = \mathbf{I}_{N_\mathrm{s}}$, and the transmit power is loaded according to $f_M^j = \sqrt{\frac{1}{\sigma_{k_j,2M}}\left(c - \frac{1}{\beta\psi_{k_j,M}}\right)^+}, 1 \le j \le N_\mathrm{s}$.

4.4 Progressive Hybrid Combining Design

In the previous section, by heuristically assuming that the fully digital linear MMSE combiner can be perfectly reconstructed by the hybrid RF-baseband counterpart, we were able to abstract the hybrid combining effect on the achievable rate, and decouple the precoding from the combining optimization. In the existing works such as [1], the reconstruction was addressed in terms of the combining structure, and was formulated as a matrix reconstruction problem. In doing this, the fully digital solution needs to be known. However, as discussed in Sect. 3.2, such knowledge might not be practically obtainable in light of the storage requirement imposed by the high-dimensional received signals across multiple ARQ rounds. Furthermore, since the matrix reconstruction formulation attempts to generate the RF and baseband combiners simultaneously, suboptimal baseband solutions are likely to result. In view of these drawbacks, we consider applying the two-step approach from the previous section to the hybrid combining design in this section.

In hopes of alleviating the storage and computational complexity at the receiver, we first seek to decrease the dimension of the received signals through RF combining, which is carried out independently across different ARQ rounds, and then combine the reduced-dimensional RF-processed signals from all the previous and current retransmissions accessible to the baseband, as illustrated in Fig. 4.1. In particular, given the hybrid RF-baseband precoders $\{\mathbf{F}^{\star}_{\mathrm{RF},m}, \mathbf{F}^{\star}_{\mathrm{BB},m}\}_{m=1}^{M}$ generated from the procedure developed in Sect. 4.3 and the RF combiners $\{\mathbf{W}^{\star}_{\mathrm{RF},m}\}_{m=1}^{M-1}$, the idea is to design the hybrid RF-baseband combiner for the current ARQ round, i.e., $\{\mathbf{W}_{\mathrm{RF},M}, \mathbf{W}_{\mathrm{BB},M}\}$, as a good approximation of the fully digital solution in terms of error performance. To this end, we consider the minimization of MSE between the transmitted signal and the aggregate received signals, which leads to the problem formulation as

$$\begin{aligned} \min_{\mathbf{W}_{\mathrm{RF},M}, \mathbf{W}_{\mathrm{BB},M}} \quad & \mathbb{E}\left[\left\| \mathbf{s} - \mathbf{W}^{\mathrm{H}}_{\mathrm{BB},M} \mathbb{W}^{\mathrm{H}}_{\mathrm{RF},M} \tilde{\mathbf{y}}_M \right\|_2^2\right] \\ \text{s.t.} \quad & \mathbf{W}_{\mathrm{RF},M} \in \mathcal{W}_{RF}, \end{aligned} \tag{4.6}$$

where $\mathcal{W}_{RF}$ denotes the set of feasible RF combiners with constant-modulus elements. It is straightforward to show that the objective in (4.6), denoted as $\mathcal{E}$, can be evaluated as

$$\begin{aligned} \mathcal{E} = 1 + \sigma_n^2 \mathrm{tr}\left[\mathbf{W}^{\mathrm{H}}_{\mathrm{BB},M} \mathbb{W}^{\mathrm{H}}_{\mathrm{RF},M} \left(\beta \mathbb{H}\mathbb{H}^{\mathrm{H}} + \mathbf{I}_{MN_{\mathrm{r}}}\right) \mathbb{W}_{\mathrm{RF},M} \mathbf{W}_{\mathrm{BB},M}\right] \\ - 2\frac{\sqrt{\rho}}{N_{\mathrm{s}}} \Re\left\{\mathrm{tr}\left(\mathbb{H}^{\mathrm{H}} \mathbb{W}_{\mathrm{RF},M} \mathbf{W}_{\mathrm{BB},M}\right)\right\}. \end{aligned} \tag{4.7}$$

In a similar manner to the hybrid precoding design, the two-step approach to the issue of hybrid combining first chooses an RF combiner from the feasible set $\mathcal{W}_{RF}$, say $\mathbf{B}_{\mathrm{RF},M}$, and then conditioned on the RF combiner $\mathbf{B}_{\mathrm{RF},M}$, computes the optimal baseband combiner $\mathbf{W}^{\star}_{\mathrm{BB},M}(\mathbf{B}_{\mathrm{RF},M})$. This procedure is repeated to generate a pair $\{\mathbf{B}_{\mathrm{RF},M}, \mathbf{W}^{\star}_{\mathrm{BB},M}(\mathbf{B}_{\mathrm{RF},M})\}$ for each chosen $\mathbf{B}_{\mathrm{RF},M} \in \mathcal{W}_{RF}$, where the MMSE-achieving pair is treated as the solution to the original problem. By solving the first-order derivative of (4.7), i.e.,

$$\frac{\partial \mathcal{E}}{\partial \mathbf{W}^{\mathrm{H}}_{\mathrm{BB},M}} = \sigma_n^2 \mathbb{W}^{\mathrm{H}}_{\mathrm{RF},M} \left(\beta \mathbb{H}\mathbb{H}^{\mathrm{H}} + \mathbf{I}_{MN_{\mathrm{r}}}\right) \mathbb{W}_{\mathrm{RF},M} \mathbf{W}_{\mathrm{BB},M} - \frac{\sqrt{\rho}}{N_{\mathrm{s}}} \mathbb{W}^{\mathrm{H}}_{\mathrm{RF},M} \mathbb{H} = \mathbf{0},$$

the baseband combiner $\mathbf{W}^{\star}_{\mathrm{BB},M}\left(\mathbf{W}_{\mathrm{RF},M}\right)$, as a function of the RF combiner, can be expressed in a closed form as

$$\mathbf{W}^{\star}_{\mathrm{BB},M}\left(\mathbf{W}_{\mathrm{RF},M}\right) = \frac{\sqrt{\rho}}{N_{\mathrm{s}}\sigma_n^2} \left[\mathbb{W}^{\mathrm{H}}_{\mathrm{RF},M} \left(\beta \mathbb{H}\mathbb{H}^{\mathrm{H}} + \mathbf{I}_{MN_{\mathrm{r}}}\right) \mathbb{W}_{\mathrm{RF},M}\right]^{-1} \mathbb{W}^{\mathrm{H}}_{\mathrm{RF},M} \mathbb{H}. \tag{4.8}$$

To facilitate the implementation of such a two-step procedure, we further reduce the infinite set of constant-modulus RF combiners $\mathcal{W}_F$ to either the finite set of $N_{\rm cl}N_{\rm ray}$ receive array response vectors $\{\mathbf{a}_{\rm r}(\phi^{\rm r}_{il})\}_{i,l}$ [1] or the $N_{\rm r}$ columns of the $N_{\rm r}$-dimensional DFT matrix $\mathcal{D}_{\rm r} = \left[\frac{1}{\sqrt{N_{\rm r}}}e^{-\frac{j2\pi mn}{N_{\rm r}}}\right], 0 \leq m, n \leq N_{\rm r} - 1$. We formally summarize the proposed strategy for MMSE hybrid combining for the Mth ARQ round as follows.

Algorithm 4.2 Two-step progressive hybrid RF-baseband combining design

1. Construct $\mathbf{B}_{{\rm RF},M}$ and accordingly $\mathbb{B}_{{\rm RF},M} \triangleq {\rm blkdiag}\{\mathbf{W}^{\star}_{{\rm RF},1}, \ldots, \mathbf{W}^{\star}_{{\rm RF},M-1}, \mathbf{B}_{{\rm RF},M}\}$ by choosing $n_{\rm r}$ vectors from either the set of $N_{\rm cl}N_{\rm ray}$ receive array response vectors $\mathcal{A}_{\rm r} = \{\mathbf{a}_{\rm r}(\phi^{\rm r}_{il})\}_{i,l}$ or the $N_{\rm r}$ columns of the DFT matrix $\mathcal{D}_{\rm r}$.
2. For the selected $\mathbf{B}_{{\rm RF},M}$, compute the MMSE baseband combiner based on (4.8) and the MSE for the pair $\left\{\mathbf{B}_{{\rm RF},M}, \mathbf{W}^{\star}_{{\rm BB},M}\left(\mathbf{B}_{{\rm RF},M}\right)\right\}$ based on (4.7).
3. Repeat step 2 for each combination of $\mathbf{B}_{{\rm RF},M}$ from $\mathcal{A}_{\rm r} = \{\mathbf{a}_{\rm r}(\phi^{\rm r}_{il})\}_{i,l}$ or from the columns of $\mathcal{D}_{\rm r}$, and choose the pair $\left\{\mathbf{B}_{{\rm RF},M}, \mathbf{W}^{\star}_{{\rm BB},M}\left(\mathbf{B}_{{\rm RF},M}\right)\right\}$ from step 2 that minimizes the MSE as the solution to (4.6).

4.5 Illustrative Results and Discussions

In this section, we present illustrative results comparing the performance of various precoding and combing methods for massive MIMO systems with packet retransmissions. The channel model is realized assuming $N_{\rm cl} = 5$ scattering clusters with $N_{\rm ray} = 2$ rays per cluster. We assume ULAs with directional antennas at the transmitter and omni-directional antennas at the receiver. The antenna elements are critically spaced, i.e., $d_\lambda = 0.5$. For each cluster, the mean azimuth AoD is assumed to be uniformly distributed over a 60° sector angle, i.e., $(0^\circ, 60^\circ)$, whereas the mean azimuth AoA at the receiver is uniformly distributed over $(0^\circ, 360^\circ)$. The azimuth AoA and AoD of each ray are Laplacian distributed with angle spread of 7.5°. We assess the two proposed schemes of progressive hybrid precoding and combining (PHPC): (1) the columns of the RF precoder and combiner are chosen from the set of array response vectors, as denoted by PHPC-AR; (2) the columns of the RF precoder and combiner are chosen from those of the DFT matrices, as denoted by PHPC-DFT. The performance of PHPC is numerically evaluated in terms of achievable rates and MSE. In the former case, we present for comparison the optimal progressive digital precoding (OPDP) scheme of [7], where an ML receiver was employed. Hence, the rate performance of such a design can be treated as a benchmark for any other precoding and combining scheme. In the latter case, we plot for comparison the OPDP scheme of [9] which addressed the minimization of MSE. The sparse precoding and combining (SPC) approach of [1] serves as another baseline. We note that SPC [1] does not take into account packet retransmissions, and it cannot be straightforwardly extended to systems equipped with packet retransmission. This is because the premise of matrix approximation that the idea of SPC was built upon would no longer hold if the previous retransmission attempts were incorporated. Hence, we adapt SPC [1] to the case of ARQ by designing the sparse precoder and combiner for each ARQ round independently. In doing this, we may gain some insights into when the use of time diversity in the precoding/combining design would become advantageous. In view of the limited scattering in the environment, only a small number of data streams is assumed to be transmitted. All the results presented here are averaged over 1000 random realizations of the massive MIMO channel.

4.5.1 *Small* M

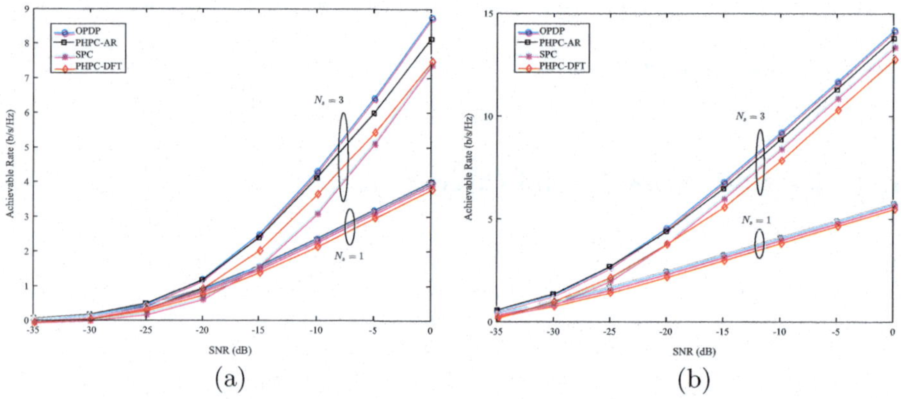

Fig. 4.2 Achievable rates of various ARQ precoding and combining schemes with $n_t = n_r = 3$, $M = 2$, and angle spread of 7.5°. **(a)** $N_t = 32$, $N_r = 8$. **(b)** $N_t = 128$, $N_r = 32$

Figure 4.2 shows a performance comparison of different precoding and combining schemes with $n_t = n_r = 3$ for the case of $M = 2$ (i.e., the occurrence of one additional retransmission) in response to different array dimensions and different numbers of data streams. The following observations can be made. For the single-stream beamforming, i.e., $N_s = 1$, all the schemes exhibit an almost indistinguishable performance. For an increased number of data streams, e.g., $N_s = 3$, the performance of the PHPC-AR method is close to that of the OPDP scheme in the low SNR regime (less than -20 dB), while suffering a slightly widening performance gap as the SNR increases. With $N_s = 3$, the SPC scheme offers a much poorer performance even at a low SNR (see Fig. 4.2a). This can be explained by the fact that SPC [1] minimizes the distance between the hybrid precoder/combiner and the fully digital optimal counterpart in one attempt without considering waterfilling-based power loading/MMSE combining at baseband. By increasing the antenna array dimensions at the transmitter and receiver, the performance loss for all the precoding and combing schemes relative to the OPDP method is reduced, as shown in Fig. 4.2b. It is noted that for a larger antenna array dimension as in Fig. 4.2b, the DFT-based PHCP method is outperformed by SPC in the regime of SNR above −20 dB. This can be attributed to the fact that PHPC-DFT only exploits the coarse-grained quantized AoDs/AoAs for RF precoding/combining while the RF precoder/combiner of SPC draws on the fine-grained information of AoDs/AoAs of all the scattering paths. In the regime of low received SNR due to (1) low beamforming gain (enabled by a not-so-large-dimensional antenna array in Fig. 4.2a and/or (2) low transmit power (SNR below −20 dB in Fig. 4.2b), power loading at baseband has a noticeable effect on the achievable rate. In this case, SPC suffers a significant performance loss from suboptimal power allocation, and is thus outperformed by PHCP-DFT. However, as the received SNR increases (above −20 dB) coupled with a substantial beamforming gain as in Fig. 4.2b, the benefit of optimal power loading tends to diminish, while the effectiveness of spatial separation between data streams in the beam domain becomes increasingly relevant to the rate performance. In this case, the performance loss caused by the quantized angular information used in the RF precoding/combining of PHCP-DFT cannot be compensated for by the diminishing performance gain from the optimal power loading at baseband. Therefore, SPC delivers a superior rate to PHCP-DFT.

4.5.2 *Large* M

Figure 4.3 shows a performance comparison with $M = 6$. With more rounds of signal retransmission, the performance degradation of the PHPC methods relative to OPDP lessens. Interestingly, by exploiting the increased time diversity, the

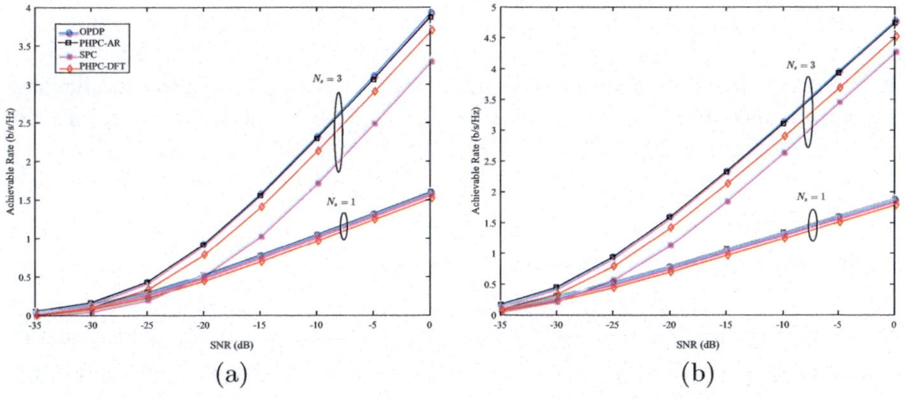

Fig. 4.3 Achievable rates of various ARQ precoding and combining schemes with $n_t = n_r = 3$, $M = 6$, and angle spread of 7.5°. **(a)** $N_t = 32$, $N_r = 8$. **(b)** $N_t = 64$, $N_r = 16$

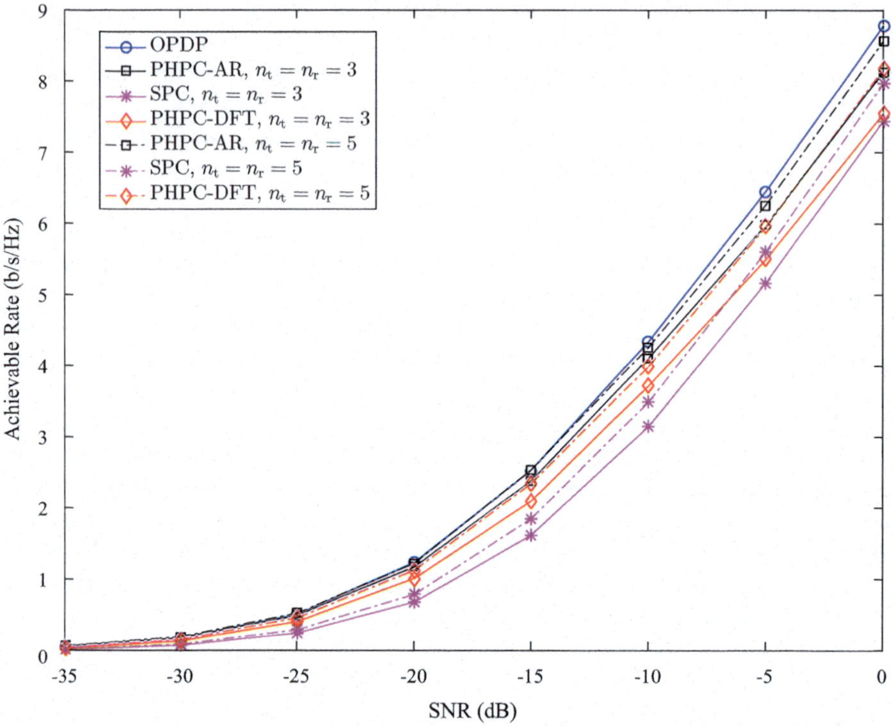

Fig. 4.4 Achievable rates of various ARQ precoding and combining schemes with different number of RF chains at the transmitter and receiver for N_t =32, N_r = 8, M = 2, N_s = 3, and angle spread of 7.5°

rate performance of PHPC-AR becomes comparable to that of OPDP even for a smaller array dimension of $N_t = 32$, $N_r = 8$ while SPC still suffers a remarkable performance degradation as the hybrid precoding and combining are optimized independently of the previous transmission attempts. In this case, one has to rely on increased spatial diversity, as provided by an increased array size, to alleviate the performance loss. This is evidenced by comparing the performance of SPC in Fig. 4.3a with that in Fig. 4.3b.

4.5.3 *Increasing Number of RF Chains*

We would like to study how the performance gap between OPDP and the PHPC schemes behaves in response to an increased number of RF chains at the transmitter and receiver. Figure 4.4 shows a performance comparison of various methods where we increase the number of RF chains available at the transmitter and receiver from $n_t = n_r = 3$ to $n_t = n_r = 5$. For the simulation, we consider an antenna configuration of $N_t = 32, N_r = 8$ with $N_s = 3$ data streams and $M = 2$ ARQ rounds. Not surprisingly, we see that the performance loss for all the methods relative to OPDP is reduced since a greater flexibility of generating hybrid precoding/combining solutions is enabled by an increased number of RF chains. In particular, PHPC-AR has a negligible performance degradation compared with the optimal OPDP scheme.

4.5.4 *Impact of Angle Spread*

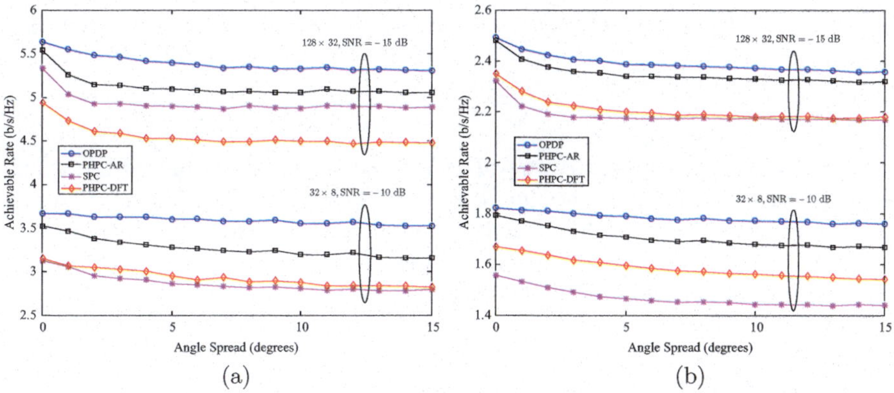

Fig. 4.5 Achievable rates versus azimuth angle spread at the transmitter and receiver for $N_s = 2$, $n_t = n_r = 2$. (**a**) $M = 2$. (**b**) $M = 6$

In Fig. 4.5, we illustrate a performance comparison of different precoding and combining methods for a varied degree of scattering typically found in a mmWave propagation environment. It is seen from Fig. 4.5a, b that the hybrid precoding/combining structure is more sensitive to increased angle spread, i.e., richer scattering, than the fully digital implementation. Intuitively, this can be attributed to the fact that the use of a limited number ($n_t = n_r = 2$) of array response/DFT beamforming vectors for the RF precoding and combining in the hybrid solutions results in an inevitable loss of channel power in the eigen-domain. Fortunately, the achievable rate of the proposed PHCP-AR remains fairly close to that of the optimal solution OPDP across the varied angle spread of interest. Furthermore, by comparing Fig. 4.5a, b, we see that PHCP-DFT outperforms SPC as the observed number of ARQ rounds increases from $M = 2$ to $M = 6$ for the array configuration of $N_t = 128, N_r = 32$. Although SPC benefits from the detailed information of AoDs/AoAs in the design of the RF precoder/combiner, it does not leverage the time diversity by ignoring the previous failed packet retransmissions. For a small number of ARQ rounds ($M = 2$), the advantage

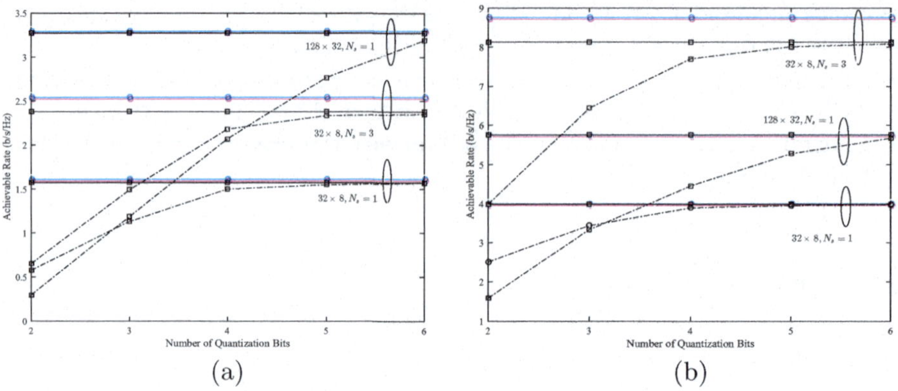

Fig. 4.6 Achievable rates versus the number of quantization bits (N_ϕ) for the azimuth AoD and AoA at the transmitter and receiver with $n_\mathrm{t} = n_\mathrm{r} = 3$, $M = 2$, and azimuth angle spread of 7.5°. (**a**) SNR $= -15$ dB. (**b**) SNR $= 0$ dB

provided by the fine-grained AoA/AoD information outweighs the performance loss from being oblivious to the time diversity. However, as the number of ARQ rounds increases ($M = 6$), the benefit of incorporating the time diversity in the precoding/combining design outweighs the loss of angle information, which leads to the superior performance of PHPC-DFT to that of SPC.

4.5.5 *Quantization of RF Precoder/Combiner*

In the proposed approach of PHCP-AR discussed in Sects. 4.3 and 4.4, we consider choosing the columns of the RF precoder $\mathbf{F}_{\mathrm{RF},M}$ and those of the RF combiner $\mathbf{W}_{\mathrm{RF},M}$ during the Mth ARQ round from the set of array response vectors $\{\mathbf{a}_\mathrm{t}(\phi^\mathrm{t}_{il})\}_{i,l}$ and $\{\mathbf{a}_\mathrm{r}(\phi^\mathrm{r}_{il})\}_{i,l}$, respectively. In practice, such knowledge of the exact AoDs $\{\phi^\mathrm{t}_{il}\}_{i,l}$ and AoAs $\{\phi^\mathrm{r}_{il}\}_{i,l}$ of all the scattering paths might not always be readily available. For example, some paths might not be spatially resolvable as a result of the finite dimension of the antenna array [10]. However, the AoD/AoA support can be inferred through estimating the mean AoD/AoA and angle spread. Suppose that the azimuth AoDs and AoAs lie within $(\phi^\mathrm{t}_{min}, \phi^\mathrm{t}_{max})$ and $(\phi^\mathrm{r}_{min}, \phi^\mathrm{r}_{max})$, respectively. One potential approach to alleviating the channel estimation overhead is to employ uniform quantization of the azimuth angular space such that the array response vectors for $\mathbf{F}_{\mathrm{RF},M}$ and $\mathbf{W}_{\mathrm{RF},M}$ during the Mth ARQ round are generated from the set of angles

$$S^a_\phi = \left\{\phi^a_{min} + \frac{k\left(\phi^a_{max} - \phi^a_{min}\right)}{2^{N_\phi} - 1}, k = 0, \ldots, 2^{N_\phi} - 1\right\}, a \in \{\mathrm{t}, \mathrm{r}\}$$

with N_ϕ denoting the number of bits used for the quantization. In Fig. 4.6, we illustrate the impact of choosing the RF precoder/combiner from a quantized angular space at the transmitter/receiver on the achievable rate. The following observations can be made. For a smaller antenna configuration of $N_t = 32, N_r = 8$, 4-bit quantization is sufficient for the single-stream beamforming. For the same antenna configuration, increasing N_s to three requires an additional bit for quantization to achieve a rate performance close to that of the perfect CSI case with complete angle information.

For a larger antenna array configuration of $N_t = 128, N_r = 32$, even for single-stream beamforming, we would need 6 bits for quantization to deliver a comparable performance with the perfect CSI case. Since there is a trade-off between the number of quantization bits to achieve a better performance and the number of combinations to be searched to generate the RF precoder/combiner, we see that $N_\phi = 5$ can be a suitable choice.

4.5.6 MSE

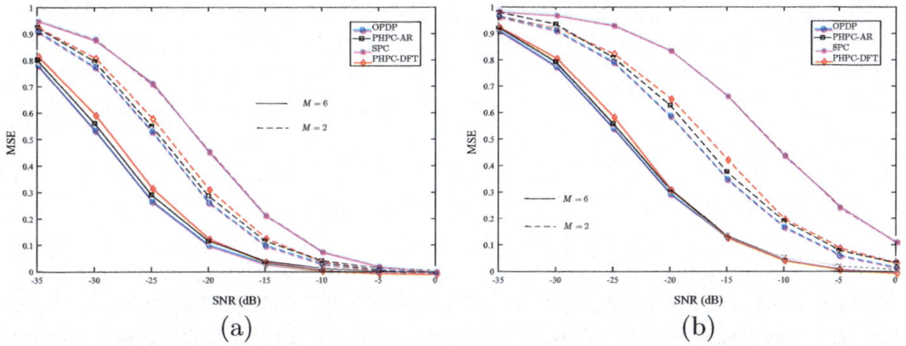

Fig. 4.7 MSE performance of various ARQ precoding and combining schemes with $n_t = n_r = 3$, and angle spread of 7.5°. **(a)** $N_s = 1$. **(b)** $N_s = 3$

In Sect. 4.4, we consider the minimization of MSE as the objective function for the design of the RF and baseband combiners for each ARQ round such that the fully digital linear MMSE combiner can be well approximated. In Fig. 4.7, we evaluate the error performance of various precoding and combining methods in terms of MSE. The OPDP curves in the MSE plots are obtained using the method proposed in [9]. As we can see from Fig. 4.7a, b, the PHPC method with both choices of RF precoders and combiners enjoys an error performance similar to that of OPDP. We also observe that the error performance of SPC is independent of the number of ARQ rounds, i.e., M, which is expected since the hybrid combiners of SPC were derived independently for each ARQ round.

4.5.7 Complexity

Both PHPC-AR and PHCP-DFT use an exhaustive search to find the desirable RF precoding and combining solutions. In the case of PHPC-AR, the RF precoding and combining involve evaluating $\binom{N_{\mathrm{cl}}N_{\mathrm{ray}}}{n_{\mathrm{t}}}$ and $\binom{N_{\mathrm{cl}}N_{\mathrm{ray}}}{n_{\mathrm{r}}}$ combinations from the set of array response vectors $\mathcal{A}_{\mathrm{t}}$ and $\mathcal{A}_{\mathrm{r}}$, respectively. In the case of PHCP-DFT where the feasible sets of the RF precoder and combiner are constrained to the columns of N_{t}-dimensional and N_{r}-dimensional DFT matrices, respectively, the number of possibilities to be evaluated is $\binom{N_{\mathrm{t}}}{n_{\mathrm{t}}}$ for the RF precoding and $\binom{N_{\mathrm{r}}}{n_{\mathrm{r}}}$ for the RF combining. If the values of N_{cl}, N_{ray}, N_{t} and N_{r} are very large, finding the solution using PHPC-AR and PHCP-DFT in real-time may not be practical. On the contrary, the quantization approach examined in Sect. 4.5.5 requires the evaluation of $\binom{2^{N_{\phi}}}{n_{\mathrm{t}}}$ and $\binom{2^{N_{\phi}}}{n_{\mathrm{r}}}$ combinations for the RF optimization at the transmitter and receiver, respectively, which are independent of the number of antennas and the number of scattering paths in the propagation environment. Using PHCP-AR and PHCP-DFT as benchmarks, one can consider the use of reduced-complexity heuristic search algorithms such as the Tabu search [11] to generate the RF solutions in real time. Once the RF precoder and combiner are found, the closed-form baseband precoder and combiner can be computed with the number of flops on the order of $\mathcal{O}\left(n_{\mathrm{t}}^3\right)$ and $\mathcal{O}\left(M^3 N_{\mathrm{r}} n_{\mathrm{r}}^2\right)$, respectively. This is a significant decrease in the computational complexity compared with the optimal solution OPDP, which requires $\mathcal{O}\left(N_{\mathrm{t}}^3\right)$ flops in the derivation of the fully digital precoder and worse-case receiver complexity scaled exponentially with N_{r}. For the case of SPC, it takes $\mathcal{O}\left(n_{\mathrm{t}}^2 N_{\mathrm{t}} N_{\mathrm{cl}} N_{\mathrm{ray}}\right)$ flops to generate the hybrid precoding solution and $\mathcal{O}\left(N_{\mathrm{cl}} N_{\mathrm{ray}} N_{\mathrm{r}}^2 n_{\mathrm{r}}\right)$ flops to generate the hybrid combining solution. For a more intuitive comparison, we list in Table 4.1 the number of real flops per ARQ round required by the dominant operations of the presented precoding and combining schemes. We focus on the case of $N_{\mathrm{s}} = 3$ data streams and $M = 6$ ARQ rounds of retransmission for both the small and large antenna configurations.

Table 4.1 Comparison of computational complexity (real flops)

	Real flops (in million)			
Antenna configuration	OPDP	PHPC-AR	PHPC-DFT	SPC
$N_{\mathrm{t}} = 32$, $N_{\mathrm{r}} = 8$	0.43	0.032	0.037	0.084
$N_{\mathrm{t}} = 128$, $N_{\mathrm{r}} = 32$	27.3	0.087	0.43	0.52

4.6 Summary

In this chapter, we considered progressive hybrid RF-baseband precoding and combining to increase the spectral efficiency by exploiting time diversity for massive MIMO with hybrid ARQ-enabled packet retransmissions. By assuming that the fully digital linear MMSE combiner can be perfectly reconstructed by the hybrid RF-baseband combiner, the development of hybrid precoding and combining solutions becomes decoupled. Toward deriving the hybrid precoder/combiner, we developed a two-step strategy for sequential joint RF-baseband optimization. Specifically, for each ARQ round, we chose the columns of the RF precoder/combiner either from the set of transmit/receive array response vectors or from the DFT-based codebooks. Conditioned on the RF precoder/combiner, we analytically derived the optimal baseband precoder/combiner. The optimal baseband precoder for the current retransmission was shown to consist of beamforming directions, which lie in the subspace spanned by the generalized eigenvectors of the effective channel and RF precoders, and power loading that depends on the precoding solutions from the previous and current retransmissions. To minimize the performance loss due to separate precoding/combining optimization, the hybrid combiner was formulated as an approximation of the linear digital combiner in terms of MSE. Illustrative results showed that the proposed progressive hybrid solutions with a limited number of RF chains provide performance improvement via exploiting the knowledge of previous ARQ retransmissions in comparison with the baseline that does not, and deliver a comparable performance with the optimal progressive digital counterpart.

Appendix 1: Proof of Theorem 4.1

We first prove that $\widetilde{\mathbf{F}}_{\mathrm{BB},M}$ is a function of the generalized eigenvectors of $\mathbf{G}_{\mathrm{RF},M}^{\mathrm{H}}\mathbf{H}_M^{\mathrm{H}}\mathbf{H}_M\mathbf{G}_{\mathrm{RF},M}$ and $\mathbf{G}_{\mathrm{RF},M}^{\mathrm{H}}\mathbf{G}_{\mathrm{RF},M}$. Consider the Cholesky factorization of $\mathbf{G}_{\mathrm{RF},M}^{\mathrm{H}}\mathbf{G}_{\mathrm{RF},M} = \mathbf{L}_M\mathbf{L}_M^{\mathrm{H}}$ and let the transformed precoder be $\overline{\mathbf{F}}_{\mathrm{BB},M} \triangleq \mathbf{L}_M^{\mathrm{H}}\widetilde{\mathbf{F}}_{\mathrm{BB},M}$. Using this transformation, we can rewrite (4.5) as

$$\begin{aligned}
&\max_{\overline{\mathbf{F}}_{\mathrm{BB},M}} \log_2\left|\widetilde{\mathbf{\Lambda}}_{M-1} + \beta\overline{\mathbf{F}}_{\mathrm{BB},M}^{\mathrm{H}}\mathbf{L}_M^{-1}\mathbf{G}_{\mathrm{RF},M}^{\mathrm{H}}\mathbf{H}_M^{\mathrm{H}}\mathbf{H}_M\mathbf{G}_{\mathrm{RF},M}\mathbf{L}_M^{-\mathrm{H}}\overline{\mathbf{F}}_{\mathrm{BB},M}\right| \\
&\text{s.t.} \quad \mathrm{tr}\left(\overline{\mathbf{F}}_{\mathrm{BB},M}^{\mathrm{H}}\overline{\mathbf{F}}_{\mathrm{BB},M}\right) = N_{\mathrm{s}},
\end{aligned} \tag{4.9}$$

where we define $\mathbf{K}_M \triangleq \mathbf{L}_M^{-1}\mathbf{G}_{\mathrm{RF},M}^{\mathrm{H}}\mathbf{H}_M^{\mathrm{H}}\mathbf{H}_M\mathbf{G}_{\mathrm{RF},M}\mathbf{L}_M^{-\mathrm{H}}$, and consider its spectral decomposition $\mathbf{K}_M = \mathbf{U}_M\mathbf{\Sigma}_M\mathbf{U}_M^{\mathrm{H}}$, with $\mathbf{U}_M$ unitary and $\mathbf{\Sigma}_M$ diagonal. Defining $\widehat{\mathbf{F}}_{\mathrm{BB},M} \triangleq \mathbf{U}_M^{\mathrm{H}}\overline{\mathbf{F}}_{\mathrm{BB},M}$, we can further simplify the problem (4.9) as

$$\begin{aligned}
&\max_{\widehat{\mathbf{F}}_{\mathrm{BB},M}} \log_2\left|\widetilde{\mathbf{\Lambda}}_{M-1} + \beta\widehat{\mathbf{F}}_{\mathrm{BB},M}^{\mathrm{H}}\mathbf{\Sigma}_M\widehat{\mathbf{F}}_{\mathrm{BB},M}\right| \\
&\text{s.t.} \quad \mathrm{tr}\left(\widehat{\mathbf{F}}_{\mathrm{BB},M}^{\mathrm{H}}\widehat{\mathbf{F}}_{\mathrm{BB},M}\right) = N_{\mathrm{s}}.
\end{aligned} \tag{4.10}$$

Using a similar line of derivation as in [7], we conclude that the optimal $\widehat{\mathbf{F}}_{\mathrm{BB},M}$ in (4.10) must be a rectangular diagonal matrix. Since $\overline{\mathbf{F}}_{\mathrm{BB},M} = \mathbf{U}_M\widehat{\mathbf{F}}_{\mathrm{BB},M}$ and $\mathbf{U}_M$ is the eigen-matrix of $\mathbf{K}_M$, it follows that $\overline{\mathbf{F}}_{\mathrm{BB},M}$ contains the eigenvectors of $\mathbf{K}_M$ with unnormalized columns. Hence, we have

$$\mathbf{L}_M^{-1}\mathbf{G}_{\mathrm{RF},M}^{\mathrm{H}}\mathbf{H}_M^{\mathrm{H}}\mathbf{H}_M\mathbf{G}_{\mathrm{RF},M}\mathbf{L}_M^{-\mathrm{H}}\overline{\mathbf{F}}_{\mathrm{BB},M} = \overline{\mathbf{F}}_{\mathrm{BB},M}\overline{\boldsymbol{\Sigma}}_M \tag{4.11}$$

where $\overline{\boldsymbol{\Sigma}}_M \in \mathbb{C}^{N_s \times N_s}$ is diagonal, and $\overline{\mathbf{F}}_{\mathrm{BB},M}$ is semi-unitary. Now substituting $\overline{\mathbf{F}}_{\mathrm{BB},M} = \mathbf{L}_M^{\mathrm{H}}\widetilde{\mathbf{F}}_{\mathrm{BB},M}$ in (4.11), we have

$$\begin{aligned}\mathbf{G}_{\mathrm{RF},M}^{\mathrm{H}}\mathbf{H}_M^{\mathrm{H}}\mathbf{H}_M\mathbf{G}_{\mathrm{RF},M}\widetilde{\mathbf{F}}_{\mathrm{BB},M} &= \mathbf{L}_M\mathbf{L}_M^{\mathrm{H}}\widetilde{\mathbf{F}}_{\mathrm{BB},M}\overline{\boldsymbol{\Sigma}}_M \\ &= \mathbf{G}_{\mathrm{RF},M}^{\mathrm{H}}\mathbf{G}_{\mathrm{RF},M}\widetilde{\mathbf{F}}_{\mathrm{BB},M}\overline{\boldsymbol{\Sigma}}_M.\end{aligned}$$

In other words, the columns of $\widetilde{\mathbf{F}}_{\mathrm{BB},M}$ consists of N_s generalized eigenvectors.

To the end of finding the exact solution for $\widetilde{\mathbf{F}}_{\mathrm{BB},M}$, let $\mathbf{A}_M \in \mathbb{C}^{n_t \times n_t}$ be the generalized eigen-matrix such that $\mathbf{G}_{\mathrm{RF},M}^{\mathrm{H}}\mathbf{H}_M^{\mathrm{H}}\mathbf{H}_M\mathbf{G}_{\mathrm{RF},M}$ and $\mathbf{G}_{\mathrm{RF},M}^{\mathrm{H}}\mathbf{G}_{\mathrm{RF},M}$ are jointly diagonalized, i.e.,

$$\begin{aligned}\mathbf{A}_M\mathbf{G}_{\mathrm{RF},M}^{\mathrm{H}}\mathbf{H}_M^{\mathrm{H}}\mathbf{H}_M\mathbf{G}_{\mathrm{RF},M}\mathbf{A}_M^{\mathrm{H}} &= \boldsymbol{\Sigma}_{1M} \\ \mathbf{A}_M\mathbf{G}_{\mathrm{RF},M}^{\mathrm{H}}\mathbf{G}_{\mathrm{RF},M}\mathbf{A}_M^{\mathrm{H}} &= \boldsymbol{\Sigma}_{2M},\end{aligned}$$

where $\boldsymbol{\Sigma}_{1M} = \mathrm{diag}\{\sigma_{1,1M}, \ldots, \sigma_{n_t,1M}\}$ and $\boldsymbol{\Sigma}_{2M} = \mathrm{diag}\{\sigma_{1,2M}, \ldots, \sigma_{n_t,2M}\}$ are diagonal. Let

$$\widetilde{\mathbf{F}}_{\mathrm{BB},M} = \mathbf{A}_M\mathbf{S}_M\mathbf{P}_{1,M}^{\mathrm{H}}\widehat{\mathbf{F}}_{\mathrm{BB},M}\mathbf{P}_{2,M}$$

be the solution. For the moment, let us assume that $\mathbf{P}_{1,M}$ and $\mathbf{P}_{2,M}$ are arbitrary permutation matrices and $\mathbf{S}_M$ is an arbitrary selection matrix which selects N_s columns of $\mathbf{G}_{\mathrm{RF},M}$. After some algebraic manipulations, we can rewrite the problem (4.5) as

$$\begin{aligned}\max_{\widehat{\mathbf{F}}_{\mathrm{BB},M}} \ & \log_2\left|\mathbf{P}_{2,M}\widetilde{\boldsymbol{\Lambda}}_{M-1}\mathbf{P}_{2,M}^{\mathrm{H}} + \beta\widehat{\mathbf{F}}_{\mathrm{BB},M}^{\mathrm{H}}\mathbf{P}_{1,M}\mathbf{S}_M^{\mathrm{H}}\boldsymbol{\Sigma}_{1M}\mathbf{S}_M\mathbf{P}_{1,M}^{\mathrm{H}}\widehat{\mathbf{F}}_{\mathrm{BB},M}\right| \\ \text{s.t.} \ & \mathrm{tr}\left(\widehat{\mathbf{F}}_{\mathrm{BB},M}^{\mathrm{H}}\mathbf{P}_{1,M}\mathbf{S}_M^{\mathrm{H}}\boldsymbol{\Sigma}_{2M}\mathbf{S}_M\mathbf{P}_{1,M}^{\mathrm{H}}\widehat{\mathbf{F}}_{\mathrm{BB},M}\right) = N_s.\end{aligned} \tag{4.12}$$

We note that the operation $\mathbf{S}_M^{\mathrm{H}}\boldsymbol{\Sigma}_{1M}\mathbf{S}_M$ selects N_s diagonal elements from the n_t diagonal elements of $\boldsymbol{\Sigma}_{1M}$, as determined by the columns of $\mathbf{S}_M$. Let the selected indices be $k_1, k_2, \ldots, k_{N_s}$. We further define $\widetilde{\boldsymbol{\Sigma}}_{jM} \triangleq \mathbf{S}_M^{\mathrm{H}}\boldsymbol{\Sigma}_{1M}\mathbf{S}_M = \mathrm{diag}\{\sigma_{k_1,jM}, \ldots, \sigma_{k_{N_s},jM}\}$, $j \in \{1, 2\}$. The operation $\mathbf{P}_{1,M}\widetilde{\boldsymbol{\Sigma}}_{jM}\mathbf{P}_{1,M}^{\mathrm{H}}$ rearranges the elements of $\widetilde{\boldsymbol{\Sigma}}_{jM}$ for $j = 1, 2$ in the order determined by $\mathbf{P}_{1,M}$. Let us denote $\widehat{\boldsymbol{\Sigma}}_{1M} \triangleq \mathbf{P}_{1,M}\widetilde{\boldsymbol{\Sigma}}_{1M}\mathbf{P}_{1,M}^{\mathrm{H}}$, $\widehat{\boldsymbol{\Sigma}}_{2M} \triangleq \mathbf{P}_{2,M}\widetilde{\boldsymbol{\Sigma}}_{2M}\mathbf{P}_{2,M}^{\mathrm{H}}$, and $\widehat{\boldsymbol{\Lambda}}_{M-1} \triangleq \mathbf{P}_{2,M}\widetilde{\boldsymbol{\Lambda}}_{M-1}\mathbf{P}_{2,M}^{\mathrm{H}}$. It is straightforward to show that (4.12) can be rewritten as

$$\max_{\widehat{\mathbf{F}}_{\mathrm{BB},M}} \ \log_2\left|\widehat{\boldsymbol{\Lambda}}_{M-1} + \beta\widehat{\mathbf{F}}_{\mathrm{BB},M}^{\mathrm{H}}\mathbf{P}_{1,M}\boldsymbol{\Psi}_M\mathbf{P}_{1,M}^{\mathrm{H}}\widehat{\mathbf{F}}_{\mathrm{BB},M}\right|$$

$$\text{s.t.} \quad \text{tr}\left(\widehat{\mathbf{F}}_{\text{BB},M}^{\text{H}}\widehat{\mathbf{F}}_{\text{BB},M}\right) = N_{\text{s}}, \tag{4.13}$$

where $\boldsymbol{\Psi}_M \triangleq \text{diag}\{\psi_{k_1,M}, \ldots, \psi_{k_{N_s},M}\}$ with $\{\psi_{m,M} \triangleq \frac{\sigma_{m,1M}}{\sigma_{m,2M}}, 1 \leq m \leq n_{\text{t}}\}$. Using a result from [7], we conclude that the optimal $\widehat{\mathbf{F}}_M$ is a diagonal matrix with diagonal elements determined by the waterfilling solution of the problem (4.13), and is given by

$$\widehat{\mathbf{F}}_{\text{BB},M} = \text{diag}\left\{f_M^1, \cdots, f_M^{N_s}\right\},$$

where $f_M^j \triangleq \sqrt{\frac{1}{\sigma_{k_j,2M}}\left(c - \frac{\widehat{\lambda}_{1,M-1}}{\beta\psi_{k_j,M}}\right)^+}, 1 \leq j \leq N_{\text{s}}$.

We now prove that the permutation matrices $\mathbf{P}_{1,M}$ and $\mathbf{P}_{2,M}$ should be chosen such that the diagonal entries in $\widetilde{\Lambda}_{M-1}$ and those in $\boldsymbol{\Psi}_M$ are paired in the reverse order. Without loss of generality, we consider the case of $N_{\text{s}} = 2$. Let $\widehat{\lambda}_{2,M-1} \geq \widehat{\lambda}_{1,M-1} \geq 0$ and $\psi_{k_2,M} \geq \psi_{k_1,M} \geq 0$, and define

$$g\left(f_M^1, f_M^2\right) \triangleq \left(\widehat{\lambda}_{1,M-1} + f_M^1\psi_{k_2,M}\right)\left(\widehat{\lambda}_{2,M-1} + f_M^2\psi_{k_1,M}\right),$$

$$g\left(f_M'^1, f_M'^2\right) \triangleq \left(\widehat{\lambda}_{1,M-1} + f_M'^2\psi_{k_1,M}\right)\left(\widehat{\lambda}_{2,M-1} + f_M'^1\psi_{k_2,M}\right),$$

with constraints $f_M^1 + f_M^2 \leq N_{\text{s}}$ and $f_M'^1 + f_M'^2 \leq N_{\text{s}}$. To prove that the reverse pairing is optimal, we need to show that for any given $f_M'^1, f_M'^2 \geq 0$, we have $g\left(f_M^1, f_M^2\right) \geq g\left(f_M'^1, f_M'^2\right)$ with $f_M^1 + f_M^2 \leq f_M'^1 + f_M'^2$. To this end, let us consider two cases:

1. $\widehat{\lambda}_{1,M-1} + f_M'^2\psi_{k_1,M} \leq \widehat{\lambda}_{2,M-1}$: In this case, the following inequality holds true, i.e.,

$$1 + \frac{\psi_{k_2,M} f_M'^1}{\widehat{\lambda}_{1,M-1} + f_M'^2\psi_{k_1,M}} + \frac{\left(\psi_{k_2,M} - \psi_{k_1,M}\right) f_M'^2}{\widehat{\lambda}_{1,M-1} + f_M'^2\psi_{k_1,M}} \geq 1 + \frac{\psi_{k_2,M} f_M'^1}{\widehat{\lambda}_{2,M-1}},$$

which can be simplified to

$$\left[\widehat{\lambda}_{1,M-1} + \psi_{k_2,M}\left(f_M'^2 + f_M'^1\right)\right]\widehat{\lambda}_{2,M-1}$$
$$\geq \left(\widehat{\lambda}_{2,M-1} + f_M'^1\psi_{k_2,M}\right)\left(\widehat{\lambda}_{1,M-1} + f_M'^2\psi_{k_1,M}\right),$$

and is seen to be equivalent to

$$g\left(f_M^1 = f_M'^1 + f_M'^2, f_M^2 = 0\right) \geq g\left(f_M'^1, f_M'^2\right).$$

2. $\widehat{\lambda}_{1,M-1} + f_M'^2\psi_{k_1,M} > \widehat{\lambda}_{2,M-1}$: In this case, by choosing

$$f_M^1 = \frac{\left(\widehat{\lambda}_{2,M-1} - \widehat{\lambda}_{1,M-1}\right)}{\psi_{k_2,M}} + f_M'^1 \text{ and } f_M^2 = \frac{\left(\widehat{\lambda}_{1,M-1} - \widehat{\lambda}_{2,M-1}\right)}{\psi_{k_1,M}} + f_M'^2,$$

we have $g\left(f_M^1, f_M^2\right) = g\left(f_M'^1, f_M'^2\right)$,and

$$f_M^1 + f_M^2 = \frac{\widehat{\lambda}_{2,M-1} - \widehat{\lambda}_{1,M-1}}{\psi_{k_2,M}} + \frac{\widehat{\lambda}_{1,M-1} - \widehat{\lambda}_{2,M-1}}{\psi_{k_1,M}} + f_M'^1 + f_M'^2 \leq f_M'^1 + f_M'^2.$$

By induction, we can extend the result to claim that $\mathbf{P}_{1,M}$ and $\mathbf{P}_{2,M}$ should be chosen such that the diagonal elements of $\widetilde{\boldsymbol{\Lambda}}_{M-1}$ and $\boldsymbol{\Psi}_M$ are paired in the reverse order. Finally, we conclude the proof by noting that since our objective is the maximization of $\left|\widehat{\boldsymbol{\Lambda}}_{M-1} + \widehat{\mathbf{F}}_{\mathrm{BB},M}^{\mathrm{H}} \widehat{\boldsymbol{\Psi}}_M \widehat{\mathbf{F}}_{\mathrm{BB},M}\right|$ in (4.13), it is trivial to see that the indices $k_1, \cdots, k_{N_s}$ of the selection matrix $\mathbf{S}_M$ (which determine the entries of $\boldsymbol{\Psi}_M$) should correspond to the largest N_s elements of the set $\{\psi_{1,M}, \ldots, \psi_{n_\mathrm{t},M}\}$.

References

1. O. El Ayach, S. Rajagopal, S. Abu-Surra, Z. Pi, and R. W. Heath, "Spatially sparse precoding in millimeter wave MIMO systems," *IEEE Trans. Wireless Commun.*, vol. 13, no. 3, pp. 1499–1513, Mar. 2014.
2. A. Liu and V. K. N. Lau, "Phase only RF precoding for massive MIMO systems with limited RF chains," *IEEE Trans. Signal Process.*, vol. 62, no. 17, pp. 4505–4515, Sep. 2014.
3. X. Yu, J.-C. Shen, J. Zhang, and K. B. Letaief, "Alternating minimization algorithms for hybrid precoding in millimeter wave MIMO systems," *IEEE J. Sel. Topics Signal Process.*, vol. 10, no. 3, pp. 485–500, Apr. 2016.
4. H. Xu, V. Kukshya, and T. Rappaport, "Spatial and temporal characteristics of 60-GHz indoor channels," *IEEE J. Sel. Areas Commun.*, vol. 20, no. 3, pp. 620–630, Apr. 2002.
5. T. L. Marzetta, "Massive MIMO: An introduction," *Bell Labs Technical Journal*, vol. 20, pp. 11–22, 2015.
6. D. P. Palomar and S. Barbarossa, "Designing MIMO communication systems: Constellation choice and linear transceiver design," *IEEE Trans. Signal Process.*, vol. 53, no. 10, pp. 3804–3818, Oct. 2005.
7. H. Sun, H. Samra, Z. Ding, and J. Manton, "Constrained capacity of linear precoded ARQ in MIMO wireless systems," in *Proc. IEEE Int. Conf. Acoustics, Speech and Signal Process.*, Philadelphia, PA, USA, Mar. 2005.
8. A. Adhikary, J. Nam, J.-Y. Ahn, and G. Caire, "Joint spatial division and multiplexing - The large-scale array regime," *IEEE Trans. Inf. Theory*, vol. 59, no. 10, pp. 6441–6463, Oct. 2013.
9. H. Sun, J. H. Manton, and Z. Ding, "Progressive linear precoder optimization for MIMO packet retransmissions," *IEEE J. Sel. Areas Commun.*, vol. 24, no. 5, pp. 448–456, Mar. 2006.
10. D. Tse and P. Viswanath, *Fundamentals of Wireless Communications*. New York, NY, USA: Cambridge University Press, 2005.
11. F. Glover and M. Laguna, *Tabu Search*. New York, NY, USA: Springer, 1997.

Chapter 5
Nonlinear Hybrid Precoding for Massive MIMO with Fractional Frequency Reuse

5.1 Introduction

In this chapter, we shift our focus from the point-to-point massive multiple-input multiple-output (MIMO) as considered in Chaps. 3 and 4 to the massive MIMO broadcast channel. Although low-complexity linear precoding schemes such as zero forcing (ZF) and regularized ZF (RZF) work sufficiently well in the presence of user diversity [1], they nonetheless tend to suffer severe power loss in a homogeneous multi-user environment, where the users experience similar channel conditions and require an equal-rate performance. This is especially the case when the system is almost fully loaded, i.e., the number of users scheduled is near the maximum number of data streams that can be physically supported [2]. In Chap. 2, it was mentioned that such a phenomenon can be effectively alleviated by vector perturbation (VP), which avoids transmission along the ill-behaving eigen-channel by introducing additional degrees of freedom in the form of an integer perturbation vector [3].

We are interested in a scenario where users are geographically clustered at multiple hotspots, and users from the same hotspot are close enough to experience the same transmit spatial correlation while separated a few wavelengths apart to experience uncorrelated small-scale channel fading. Built upon the aforementioned insights, the nonlinear minimum mean square error (MMSE)-VP technique might prove useful for system capacity maximization in this case. This motivates us to explore the use of the MMSE-VP technique in a two-timescale hybrid RF-baseband precoding design. In particular, the RF precoder is updated based on the statistical channel state information (CSI), and strives to achieve good spatial separation between user clusters. Depending on the severity of the residual inter-cluster interference, instantaneous effective CSI-based MMSE-VP precoding can be performed either jointly across all user clusters or separately on each cluster at

T. Le-Ngoc, R. Mai, *Hybrid Massive MIMO Precoding in Cloud-RAN*,
Wireless Networks, https://doi.org/10.1007/978-3-030-02158-0_5

baseband. The joint MMSE-VP offers the best performance at the cost of increased computational complexity and channel estimation overhead while the cluster-wise MMSE-VP strikes a performance-complexity trade-off. Considering both variations of MMSE-VP at baseband, RF precoder designs with respect to nonuniform- and constant-modulus elements are addressed.

Although the existing techniques of nonlinear precoding in the literature, such as [4–6], employ a two-stage processing structure, they assume perfect CSI, and cannot be directly applied to the case where RF processing is adaptive to statistical CSI. In other words, a novel design strategy is needed. In this chapter, we propose a two-step solution technique, which enables non-iterative RF-baseband optimization. In particular, perfect effective CSI-based MMSE-VP precoding is first derived for the baseband, built upon which we formulate the statistical CSI-based RF designs as solutions to stochastic optimization problems subject to orthonormal constraints. As a result of the nonlinear nature of VP, there is no closed-form characterization of the objective functions. Nonetheless, we analytically prove that for the single-cluster scenarios, statistical eigen-beamforming in the RF domain is indeed an optimal solution. To facilitate numerical optimization for the multi-cluster scenarios, we propose mathematically tractable lower bounds as approximations of the objective functions. By viewing the orthonormality-constrained nonuniform-modulus RF precoder as a point on matrix manifolds, the lower bounds are numerically optimized by trust-region Newton algorithms on Riemannian manifolds, which achieve global convergence at a locally superlinear rate. Moreover, the problem of discrete Fourier transform (DFT) beam selection in the RF domain is addressed, which serves as an effective alternative to the constant-modulus RF precoder design. Recognizing that the objective function can be expressed as a difference of increasing functions, we develop branch-reduce-and-bound (BRnB) techniques to find a global optimum with reduced computational complexity. Through numerical simulations, we demonstrate that for single-cluster transmission, introduction of VP offsets the performance penalty incurred by limited RF chains and partial CSI in the RF domain in comparison with the fully digital linear RZF solutions. For multi-cluster transmission, we illustrate the superiority of the proposed approach for *joint* RF-baseband design to the other state-of-the-art baselines.

The rest of this chapter is organized as follows. In Sect. 5.2, the system model and problem statement of hybrid precoding with MMSE-VP for massive MIMO downlink are presented. We develop the proposed solution techniques for nonlinear hybrid precoding in Sect. 5.3 and for nonlinear hybrid precoding with RF phase-shifting in Sect. 5.4. Illustrative results are provided in Sect. 5.5, where the error performance of the proposed nonlinear hybrid precoding solutions is numerically compared with other state-of-the-art hybrid schemes. Concluding remarks are made in Sect. 5.6.

5.2 System Model and Problem Statement

As illustrated in Fig. 5.1, let us consider a multi-user massive MIMO system, where a base station (BS) equipped with a uniform linear array (ULA) of N_t elements serves K single-antenna users on the downlink.[1] On account of limitations of

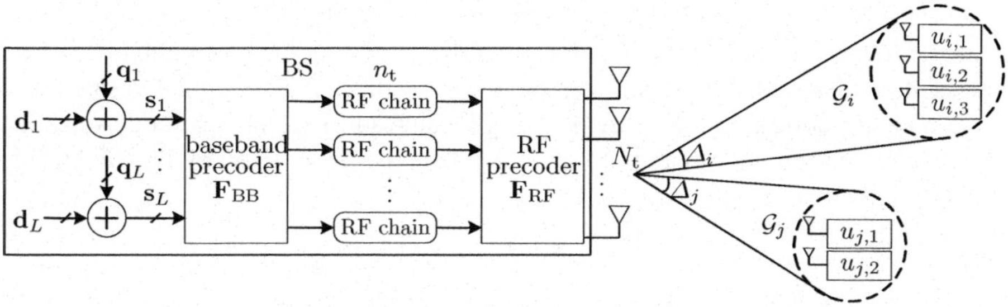

Fig. 5.1 Hybrid precoding with VP for multiple clusters of users on a massive MIMO downlink where the perfect RF-precoded CSI is available at baseband and the channel spatial correlation is known at RF

hardware complexity and power consumption, the large-scale ULA is driven by n_t RF chains, satisfying $N_\mathrm{t} \gg n_\mathrm{t} \geq K$. Suppose that the users are geographically partitioned into L non-overlapping clusters $\mathcal{G}_1, \ldots, \mathcal{G}_L$ with $|\mathcal{G}_i| = g_i$, $\sum_{i=1}^{L} g_i = K$, and $\mathcal{G}_i \cap \mathcal{G}_j = \emptyset, \forall i \neq j$. Let $u_{i,k}$, $k = 1, \ldots, g_i$, $i = 1, \ldots, L$ denote the kth user of $\mathcal{G}_i$. We assume a block-fading channel with a block length of T channel uses. The received signal by user-$u_{i,k}$ for the tth channel use reads as

$$y_{i,k}[t] = \mathbf{h}_{i,k}^{[n]\mathrm{H}} \mathbf{x}[t] + n_{i,k}[t],$$

or collectively for $\mathcal{G}_i$ as

$$\underbrace{\left[y_{i,1}[t], \ldots, y_{i,g_i}[t]\right]^\mathrm{T}}_{\triangleq \mathbf{y}_i[t]} = \underbrace{\left[\mathbf{h}_{i,1}^{[n]}, \ldots, \mathbf{h}_{i,g_i}^{[n]}\right]^\mathrm{H}}_{\triangleq \mathbf{H}_i^{[n]}} \mathbf{x}[t] + \underbrace{\left[n_{i,1}[t], \ldots, n_{i,g_i}[t]\right]^\mathrm{T}}_{\triangleq \mathbf{n}_i[t]},$$

[1]In this work, we consider the use of ULA for transmit beamforming in the azimuth plane for the simplicity of solution development. In the case of uniform rectangular arrays, which are more likely to be found in practical deployment of massive MIMO, the transmit spatial correlation can be approximately written as a Kronecker product of the correlation in the azimuth and elevation planes [7]. This implies that the design of full-dimensional beamforming in the RF domain can be decomposed as azimuth beamforming and elevation beamforming, which allows for a direct application of the solution techniques developed in the subsequent presentation to the azimuth and elevation planes, respectively.

where $t \in \mathcal{T}_n \triangleq \{t' \in \mathbb{Z} | (n-1)T < t' \leq nT\}$ is the tth channel use of the nth channel fading block $\mathcal{T}_n$, $\mathbf{x}[t] \in \mathbb{C}^{N_\mathrm{t}}$ represents the transmit signal, $\mathbf{h}_{i,k}^{[n]} \in \mathbb{C}^{N_\mathrm{t}}$ is the channel vector between the BS and user-$u_{i,k}$, and $n_{i,k}[t] \sim \mathcal{CN}(0, \sigma_n^2)$ is the spatially white additive Gaussian noise with variance σ_n^2. In particular, the spatially correlated channel is modeled as $\mathbf{h}_{i,k}^{[n]} = \boldsymbol{\Sigma}_i^{1/2}\mathbf{h}_{i,k}^{w[n]}$, where $\boldsymbol{\Sigma}_i = \mathbb{E}[\mathbf{h}_{i,k}^{[n]}\mathbf{h}_{i,k}^{[n]\mathrm{H}}]$ is the transmit correlation matrix, and $\mathbf{h}_{i,k}^{w[n]} \sim \mathcal{CN}(\mathbf{0}, \mathbf{I}_{N_\mathrm{t}})$ represents the small-scale Rayleigh fading channel. Without loss of generality, we consider transmission over the nth channel block, and omit the channel block index unless otherwise stated. Note that users in the same cluster are assumed to share the same transmit correlation. For the one-ring local scattering, the spatial correlation coefficient between the qth and the pth antenna at the BS is given by Shiu et al. [8] and Adhikary et al. [9]

$$[\boldsymbol{\Sigma}_i]_{p,q} = \frac{1}{\Delta_i}\int_{\bar{\theta}_i-\frac{\Delta_i}{2}}^{\bar{\theta}_i+\frac{\Delta_i}{2}} e^{\jmath 2\pi d_\lambda (p-q)\sin(\theta)}d\theta,$$

where d_λ is the inter-antenna spacing normalized with respect to the wavelength, $\bar{\theta}_i$ is the mean azimuth angle of departure (AoD) of scatterers and Δ_i is the angle spread. Based on the spectral decomposition $\boldsymbol{\Sigma}_i = \mathbf{U}_i\boldsymbol{\Lambda}_i\mathbf{U}_i^\mathrm{H}$ with $\mathbf{U}_i$ unitary and $\boldsymbol{\Lambda}_i$ diagonal, we define $\boldsymbol{\Sigma}_i^{1/2} \triangleq \mathbf{U}_i\boldsymbol{\Lambda}_i^{1/2}$, i.e., $\boldsymbol{\Sigma}_i = \boldsymbol{\Sigma}_i^{1/2}(\boldsymbol{\Sigma}_i^{1/2})^\mathrm{H}$. Let r_i denote the number of significant eigenvalues of $\boldsymbol{\Sigma}_i$. Generally, $r_i \ll N_\mathrm{t}$ as a result of limited local scattering, i.e., small Δ_i. Besides, it can be seen that $\boldsymbol{\Sigma}_i$ is a Toeplitz matrix. Such an observation will serve as a basis for DFT beam selection as a simplified constant-modulus design approach in Sects. 5.3.3 and 5.4.3.

The transmit signal is expressed as

$$\mathbf{x}[t] = \sum_{i=1}^{L} \mathbf{F}_{\mathrm{RF},i}\mathbf{x}_{\mathrm{BB},i}[t],$$

where $\mathbf{F}_{\mathrm{RF},i} \in \mathbb{C}^{N_\mathrm{t}\times m_i}$ with $m_i \geq g_i$ is the RF analog precoder that adaptively steers an m_i-dimensional RF beam-space for the coverage of $\mathcal{G}_i$, and $\mathbf{x}_{\mathrm{BB},i}[t] \in \mathbb{C}^{m_i}$ is the baseband precoded signal. The RF analog precoders, written collectively as $\mathbf{F}_\mathrm{RF} \triangleq [\mathbf{F}_{\mathrm{RF},1}, \ldots, \mathbf{F}_{\mathrm{RF},L}]$, are adaptive to the global statistical CSI $\{\boldsymbol{\Sigma}_i\}_{i=1}^L$, and satisfy orthonormality to avoid strong intra- and inter-cluster interference, i.e., $\mathbf{F}_\mathrm{RF}^\mathrm{H}\mathbf{F}_\mathrm{RF} = \mathbf{I}_{n_\mathrm{t}}$. For analytical simplicity, the average power constraint over a transmission block of T channel uses is imposed, i.e., $\frac{1}{T}\sum_{t\in\mathcal{T}_n}\|\mathbf{x}[t]\|^2 = \frac{1}{T}\sum_{t\in\mathcal{T}_n}\sum_{i=1}^{L}\|\mathbf{x}_{\mathrm{BB},i}[t]\|^2 \leq P_\mathrm{t}$.

To generate $\{\mathbf{x}_{\mathrm{BB},i}[t]\}_{i=1}^L$ through VP for the tth channel use, the original data symbol $\mathbf{d}_i[t]$ for $\mathcal{G}_i$, $1 \leq i \leq L$, is first perturbed with a complex integer vector $\mathbf{q}_i[t] \in \mathbb{Z}^{g_i} + \jmath\mathbb{Z}^{g_i}$, and the data symbol $\mathbf{d}[t] \triangleq [\mathbf{d}_1[t]^\mathrm{T}, \ldots, \mathbf{d}_L[t]^\mathrm{T}]^\mathrm{T}$ across all L clusters is then linearly precoded by $\mathbf{F}_\mathrm{BB} \in \mathbb{C}^{n_\mathrm{t}\times K}$. In particular, let $\mathbf{s}_i[t] = \mathbf{d}_i[t] + \tau\mathbf{q}_i[t]$ denote the perturbed signal for $\mathcal{G}_i$. The precoded signal at baseband is then given as

$$\underbrace{\left[\mathbf{x}_{\mathrm{BB},1}[t]^{\mathrm{T}},\ldots,\mathbf{x}_{\mathrm{BB},L}[t]^{\mathrm{T}}\right]^{\mathrm{T}}}_{\triangleq \mathbf{x}_{\mathrm{BB}}[t]} = \mathbf{F}_{\mathrm{BB}}\underbrace{\left[\mathbf{s}_1[t]^{\mathrm{T}},\ldots,\mathbf{s}_L[t]^{\mathrm{T}}\right]^{\mathrm{T}}}_{\triangleq \mathbf{s}[t]}.$$

Some comments on VP are in order. In view of the integer-valued $\mathbf{q}_i[t]$, the perturbation can be interpreted as adding an integer multiple of τ to each element of $\mathbf{d}_i[t]$. This implies that modulo-τ decoding, defined element-wise as $f_\tau(\omega) \triangleq \omega - \left\lfloor \frac{\omega}{\tau} + 0.5 \right\rfloor \tau$, can be used to remove the perturbation effect such that $\mathbf{d}_i[t] = f_\tau(\Re\{\mathbf{s}_i[t]\}) + \jmath f_\tau(\Im\{\mathbf{s}_i[t]\})$. From the definition of $f_\tau(\cdot)$, it is seen that correct modulo decoding is achievable provided that $-0.5\tau \le \Re\{\mathbf{d}_i[t]\} < 0.5\tau$, and $-0.5\tau \le \Im\{\mathbf{d}_i[t]\} < 0.5\tau$. Let $c_{\max}$ denote the absolute value of the largest magnitude per real dimension in the constellation diagram. On one hand, if $\tau < 2c_{\max}$, such a condition does not hold for all $\mathbf{d}_i[t]$, and thus $\mathbf{d}_i[t]$ cannot always be correctly recovered. On the other hand, when τ is made too large, optimization of the perturbation vector $\mathbf{q}_i[t]$ is likely to lead to the all-zero vector, and becomes independent of $\mathbf{d}_i[t]$ [3]. As shown in [10], a good rule of thumb is to choose $\tau = 4c_{\min}$ where $c_{\min}$ is the minimum distance of the constellation symbols per real dimension, and it is generally assumed that the same τ is used across all users for simplicity.

With scalar equalization β^{-1}, the signal estimate $\hat{\mathbf{s}}_i[t]$ for $\mathcal{G}_i$ can be expressed as

$$\hat{\mathbf{s}}_i[t] = \beta^{-1}\mathbf{H}_i^{\mathrm{H}}\mathbf{F}_{\mathrm{RF}}\mathbf{x}_{\mathrm{BB}}[t] + \beta^{-1}\mathbf{n}_i[t],$$

where $\mathbf{H}_i \triangleq \boldsymbol{\Sigma}_i^{1/2}\mathbf{H}_{i,w}$ with $\mathbf{H}_{i,w} \triangleq [\mathbf{h}_{i,1}^{w},\ldots,\mathbf{h}_{i,g_i}^{w}]$, $1 \le i \le L$. The conditional mean square error (MSE) for $\mathcal{G}_i$ at the tth channel use is defined as [10]

$$\begin{aligned}\mathcal{E}_i[t] &\triangleq \mathbb{E}_{\mathbf{n}_i[t]}\left[\left\|\mathbf{s}_i[t] - \hat{\mathbf{s}}_i[t]\right\|^2 \middle| \mathbf{d}_i[t]\right]\\ &= \beta^{-2}\left(\mathbf{x}_{\mathrm{BB}}[t]^{\mathrm{H}}\mathbf{F}_{\mathrm{RF}}^{\mathrm{H}}\mathbf{H}_i\mathbf{H}_i^{\mathrm{H}}\mathbf{F}_{\mathrm{RF}}\mathbf{x}_{\mathrm{BB}}[t] + g_i\sigma_n^2\right)\\ &\quad - 2\beta^{-1}\Re\left\{\mathbf{s}_i[t]^{\mathrm{H}}\mathbf{H}_i^{\mathrm{H}}\mathbf{F}_{\mathrm{RF}}\mathbf{x}_{\mathrm{BB}}[t]\right\} + \|\mathbf{s}_i[t]\|^2.\end{aligned} \tag{5.1}$$

Therefore, the nonlinear hybrid precoder is posed as a solution minimizing the sum MSE across all L clusters per channel use, i.e.,

$$\begin{aligned}&\min_{\mathbf{F}_{\mathrm{RF}}}\ \mathbb{E}_{\{\mathbf{H}_{i,w},\mathbf{d}_i[t]\}}\left[\min_{\{\mathbf{x}_{\mathrm{BB}}[t]\}_t,\{\mathbf{q}_i[t]\}_{i,t}} \frac{1}{T}\sum_{i=1}^{L}\sum_{t\in\mathcal{T}_n}\mathcal{E}_i[t]\right]\\ &\text{s.t.}\ \ \mathbf{F}_{\mathrm{RF}}^{\mathrm{H}}\mathbf{F}_{\mathrm{RF}} = \mathbf{I}_{n_t},\ \frac{1}{T}\sum_{t\in\mathcal{T}_n}\|\mathbf{x}[t]\|^2 \le P_t.\end{aligned} \tag{5.2}$$

Because of the nonlinear nature of VP, the statistics of the perturbed signals $\{\mathbf{s}_i[t]\}_{i=1}^{L}$ are generally not known. In order to circumvent this difficulty, the baseband precoder is obtained by instead minimizing the conditional MSE $\{\mathcal{E}_i[t]\}_{i=1}^{L}$, where only the noise is averaged out. Expectation over the channel $\{\mathbf{H}_{i,w}\}_{i=1}^{L}$ is taken since the RF precoder $\mathbf{F}_{\mathrm{RF}}$ is constrained in the subspace spanned by the spatial correlation matrices $\{\boldsymbol{\Sigma}_i\}_{i=1}^{L}$ while expectation over the data symbols $\{\mathbf{d}_i[t]\}_{i,t}$ is motivated by seeking non-iterative solution techniques. The problem statement indicates that we address the issue of hybrid precoder design by addressing first the baseband stage and then the RF stage. As shown in what follows, in doing this, the objective function can be simplified by eliminating the baseband design variables, and allows us to develop mathematically tractable approximation in the RF design. We consider two types of baseband processing: Joint-cluster processing (JCP) which performs nonlinear VP precoding across all L clusters, and per-cluster processing (PCP) which is concerned with cluster-wise nonlinear VP precoding. By taking into account self-transmission and inter-cluster interference, JCP offers the best performance. PCP, on the other hand, aims for a trade-off between performance and complexity by relying on effective separation of clusters in the RF beam domain.

5.3 Joint Hybrid MMSE-VP Precoding

In this section, we first briefly revisit the MMSE-VP precoding solution for the baseband in an attempt to simplify the objective function in (5.2). We then develop solution techniques to derive RF precoders with nonuniform- and constant-modulus elements, respectively.

5.3.1 MSE-Based Problem Formulation

In the case of JCP, a linear precoder $\mathbf{F}_{\mathrm{BB}}$ together with the perturbation vector $\{\mathbf{q}_i[t]\}_{i=1}^{L}$ are jointly derived across all L user clusters. The sum MSE for the tth channel use, conditioned on $\{\mathbf{d}_i[t]\}_{i=1}^{L}$, is rewritten as

$$\begin{aligned}\mathcal{E}_{\mathrm{jcp}}[t] &\triangleq \sum_{i=1}^{L} \mathcal{E}_i[t]\\ &= \beta^{-2}\left(\mathbf{x}_{\mathrm{BB}}[t]^{\mathrm{H}}\mathbf{F}_{\mathrm{RF}}^{\mathrm{H}}\mathbf{H}\mathbf{H}^{\mathrm{H}}\mathbf{F}_{\mathrm{RF}}\mathbf{x}_{\mathrm{BB}}[t] + K\sigma_n^2\right)\\ &\quad - 2\beta^{-1}\Re\left\{\mathbf{s}[t]^{\mathrm{H}}\mathbf{H}^{\mathrm{H}}\mathbf{F}_{\mathrm{RF}}\mathbf{x}_{\mathrm{BB}}[t]\right\} + \|\mathbf{s}[t]\|^2, \qquad (5.3)\end{aligned}$$

where $\mathbf{H} \triangleq [\mathbf{H}_1, \ldots, \mathbf{H}_L] = \boldsymbol{\Sigma}^{1/2}\mathbf{H}_w$ with $\boldsymbol{\Sigma}^{1/2} \triangleq [\boldsymbol{\Sigma}_1^{1/2}, \ldots, \boldsymbol{\Sigma}_L^{1/2}]$ and $\mathbf{H}_w \triangleq \mathrm{diag}\{\mathbf{H}_{1,w}, \ldots, \mathbf{H}_{L,w}\}$. By solving the Karush-Kuhn-Tucker (KKT) conditions, the optimal baseband solution can be shown as [10]

$$\mathbf{x}_{\mathrm{BB}}^{\star}[t] = \beta^{\star}\mathbf{F}_{\mathrm{BB}}^{\star}\mathbf{s}[t],$$

where $\mathbf{F}_{\mathrm{BB}}^{\star} = \mathbf{F}_{\mathrm{RF}}^{\mathrm{H}}\mathbf{H}(\mathbf{H}^{\mathrm{H}}\mathbf{F}_{\mathrm{RF}}\mathbf{F}_{\mathrm{RF}}^{\mathrm{H}}\mathbf{H} + \gamma^{-1}\mathbf{I}_K)^{-1}$ is the RZF precoder with the regularization coefficient $\gamma = \frac{P_{\mathrm{t}}}{K\sigma_n^2}$, and $\beta^{\star} = \sqrt{\frac{P_{\mathrm{t}}}{\frac{1}{T}\sum_{t\in\mathcal{T}_n}\|\mathbf{F}_{\mathrm{BB}}^{\star}\mathbf{s}[t]\|^2}}$ is the scaling factor yielding full power transmission.

Accordingly, the conditional MSE in (5.3) is simplified as [10]

$$\mathcal{E}_{\mathrm{jcp}}[t] = \left(\mathbf{d}[t] + \tau\mathbf{q}[t]\right)^{\mathrm{H}}\mathbf{B}^{-1}\left(\mathbf{d}[t] + \tau\mathbf{q}[t]\right),$$

where $\mathbf{B} \triangleq \mathbf{I}_K + \gamma\mathbf{H}^{\mathrm{H}}\mathbf{F}_{\mathrm{RF}}\mathbf{F}_{\mathrm{RF}}^{\mathrm{H}}\mathbf{H}$. The optimal perturbation vector is posed as the solution to the following integer least squares (LS) problem, i.e., for $t \in \mathcal{T}_n$,

$$\begin{aligned}\mathbf{q}^{\star}[t] &\triangleq \underset{\mathbf{q}[t]\in\mathbb{Z}^K+\jmath\mathbb{Z}^K}{\arg\min}\ \left(\mathbf{d}[t] + \tau\mathbf{q}[t]\right)^{\mathrm{H}}\mathbf{B}^{-1}\left(\mathbf{d}[t] + \tau\mathbf{q}[t]\right)\\ &= \underset{\mathbf{q}[t]\in\mathbb{Z}^K+\jmath\mathbb{Z}^K}{\arg\min}\ \|-\mathbf{L}\mathbf{d}[t] - \tau\mathbf{L}\mathbf{q}[t]\|^2,\end{aligned} \tag{5.4}$$

where the lower triangular matrix $\mathbf{L}$ is obtained through Cholesky decomposition of $\mathbf{B}^{-1}$, i.e., $\mathbf{B}^{-1} = \mathbf{L}^{\mathrm{H}}\mathbf{L}$. Since $\mathbf{q}[t]$ is a complex integer vector by assumption, the search for the optimal perturbation vector can be interpreted as finding the closest point to $-\mathbf{L}\mathbf{d}[t]$ in the lattice defined by $\tau\mathbf{L}$. An efficient search generally consists of two stages: Lenstra-Lenstra-Lovász (LLL) basis reduction and Schnorr-Euchner enumeration. For further details, we refer the interested reader to [11, 12]. Let $\mathcal{E}_{\mathrm{jcp}}^{\star[t]}(\mathbf{F}_{\mathrm{RF}})$ denote the resulting value from $\mathbf{q}^{\star}[t]$. The problem of hybrid precoding (5.2) is thus reduced to the following stochastic optimization problem of RF design

$$\begin{aligned}&\min_{\mathbf{F}_{\mathrm{RF}}}\mathbb{E}_{\{\mathbf{H}_{i,w},\mathbf{d}_i[t]\}}\left[\frac{1}{T}\sum_{t\in\mathcal{T}_n}\mathcal{E}_{\mathrm{jcp}}^{\star[t]}\left(\mathbf{F}_{\mathrm{RF}}\right)\right]\\ &\text{s.t. } \mathbf{F}_{\mathrm{RF}}^{\mathrm{H}}\mathbf{F}_{\mathrm{RF}} = \mathbf{I}_{n_{\mathrm{t}}}.\end{aligned} \tag{5.5}$$

Note that a feasible solution $\mathbf{F}_{\mathrm{RF}}$ to (5.5) is invariant under unitary transformation. In other words, the solution is essentially an n_{t}-dimensional subspace spanned by the columns of $\mathbf{F}_{\mathrm{RF}}$, denoted by $\lfloor\mathbf{F}_{\mathrm{RF}}\rfloor$. Under the orthonormality constraint $\mathbf{F}_{\mathrm{RF}}^{\mathrm{H}}\mathbf{F}_{\mathrm{RF}} = \mathbf{I}_{n_{\mathrm{t}}}$, the subspace can be viewed as a point on the complex Grassmann manifold $\mathcal{M}_{\mathrm{gr}} \triangleq \{\lfloor\mathbf{F}_{\mathrm{RF}}\rfloor \,|\, \mathbf{F}_{\mathrm{RF}} \in \mathbb{C}^{N_{\mathrm{t}}\times n_{\mathrm{t}}},\ \mathbf{F}_{\mathrm{RF}}^{\mathrm{H}}\mathbf{F}_{\mathrm{RF}} = \mathbf{I}_{n_{\mathrm{t}}}\}$. Generally, in order to solve (5.5), a closed-form expression for the objective is required. However, this is non-trivial due to the nonlinear perturbation search. In Sect. 5.3.2.1, we first study the special scenario where transmission takes place for a single cluster of users, and an optimal solution can be derived without explicitly evaluating the objective. In Sect. 5.3.2.2, the general case of more than one cluster of users is examined where

$\mathbb{E}_{\{\mathbf{H}_{i,w},\mathbf{d}_i[t]\}_i}[\mathcal{E}_{\text{jcp}}^{\star[t]}(\mathbf{F}_{\text{RF}})]$, $t \in \mathcal{T}_n$, is approximated by a closed-form lower bound and numerically optimized.

5.3.2 *Statistical CSI-Based RF Precoding Design*

5.3.2.1 Single-Cluster Transmission

Clearly, a feasible solution $\mathbf{F}_{\text{RF}}^{\star}$ must achieve optimality if $\mathcal{E}_{\text{jcp}}^{\star[t]}(\mathbf{F}_{\text{RF}}^{\star}) \leq \mathcal{E}_{\text{jcp}}^{\star[t]}(\mathbf{F}_{\text{RF}})$ holds for arbitrary $\mathbf{d}[t]$ and $\mathbf{H}$. In the special case where only one cluster of users, say $\mathcal{G}_i$, is scheduled for transmission, eigen-beamforming satisfies such a stronger condition.

Theorem 5.1 *Suppose that the ith cluster of users is scheduled for transmission. The optimal RF precoder $\mathbf{F}_{\text{RF}}^{\star}$ minimizing the average MSE in (5.5) is given by the subspace spanned by the eigenvectors of the transmit correlation matrix $\boldsymbol{\Sigma}_i$ associated with the largest n_{t} eigenvalues.*

Proof See Appendix 1.

5.3.2.2 Multi-Cluster Transmission

When more than one cluster of users are scheduled, RF eigen-beamforming, unfortunately, is no longer optimal. Because the probability distribution of the perturbation vector is not available as a result of the nonlinear search, an exact closed-form expression for the objective in (5.5) cannot be computed. Interestingly, by approximating the resulting MSE as uniformly distributed, a mathematically tractable lower bound can be established [13].

Lemma 5.1 *Suppose that the constellation symbols are uniformly distributed within a hyper-rectangle centered at the origin. Then the resulting MSE $\mathcal{E}_{\text{jcp}}^{\star[t]}(\mathbf{F}_{\text{RF}})$ from VP is uniformly distributed, and its expectation can be lower-bounded by*

$$\mathbb{E}_{\{\mathbf{H}_{i,w},\mathbf{d}_i[t]\}}\left[\mathcal{E}_{\text{jcp}}^{\star[t]}(\mathbf{F}_{\text{RF}})\right] \geq \underbrace{\frac{K\tau^2}{\pi e}\left\{\mathbb{E}_{\{\mathbf{H}_{i,w}\}}\left[\det\left(\mathbf{I}_K + \gamma\mathbf{H}^{\text{H}}\mathbf{F}_{\text{RF}}\mathbf{F}_{\text{RF}}^{\text{H}}\mathbf{H}\right)\right]\right\}^{-\frac{1}{K}}}_{\triangleq\, \mathcal{E}_{\text{jcp,lb}}(\mathbf{F}_{\text{RF}})}.$$

Proof See Appendix 2.

Remark 5.1 For the assumption on the constellation symbols to be valid approximately in practice, the signal constellation needs to be square and symmetric around the origin, which is satisfied by square QAM modulation, e.g., 16-, 64-, 256-QAM as defined by 3GPP Long-Term Evolution (LTE) cellular systems. Furthermore, the modulation order for the users needs to be identical. This holds naturally for equal-rate clustered users.

Remark 5.2 The expectation $\mathbb{E}_{\{\mathbf{d}_i[t]\}}[\mathcal{E}_{\text{jcp}}^{\star[t]}(\mathbf{F}_{\text{RF}})]$ can be interpreted as the second moment of the error vector $\boldsymbol{\eta}[t] \triangleq -\mathbf{L}\mathbf{d}[t] - \tau\mathbf{L}\mathbf{q}^{\star}[t]$ which lies in the Voronoi region of the lattice $\tau\mathbf{L}$ associated with the origin, denoted as $\mathcal{V}_0$. By assuming $\mathbf{d}[t]$ as uniformly distributed, $\boldsymbol{\eta}[t]$ becomes uniformly distributed within $\mathcal{V}_0$. Thus, the second moment of $\boldsymbol{\eta}[t]$ can be related to the normalized second moment (NSM) of $\mathcal{V}_0$ scaled by its volume per real dimension $\text{vol}(\mathcal{V}_0)^{\frac{1}{2K}}$. As $\mathcal{V}_0$ becomes increasingly sphere-like with $K \to \infty$, the NSM of $\mathcal{V}_0$ is asymptotically lower bounded by $\frac{1}{2\pi e}$. This observation serves as the basis for deriving the lower bound in Lemma 5.1.

For mathematical tractability, we study the lower bound in lieu of the original objective function. Thus, design of RF precoding subject to orthonormality is equivalently recast as

$$\mathbf{F}_{\text{RF}}^{\star} \triangleq \underset{\mathbf{F}_{\text{RF}} \in \mathcal{M}_{\text{gr}}}{\arg\max}\ \mathbb{E}_{\{\mathbf{H}_{i,w}\}}\left[\det\left(\mathbf{I}_K + \gamma\mathbf{H}^{\text{H}}\mathbf{F}_{\text{RF}}\mathbf{F}_{\text{RF}}^{\text{H}}\mathbf{H}\right)\right]. \tag{5.6}$$

Let $\mathcal{J}(\mathbf{F}_{\text{RF}})$ denote the objective function of (5.6). It turns out that the expectation in (5.6) can be explicitly computed, as summarized in the following theorem.

Theorem 5.2 *The objective function $\mathcal{J}(\mathbf{F}_{\text{RF}})$ can be evaluated to a closed form, which is given by*

$$\mathcal{J}(\mathbf{F}_{\text{RF}}) = \sum_{k=0}^{K} \gamma^k \sum_{\mathcal{I}_k = \bigcup_{i=1}^{L} \mathcal{I}_{i,k} \subset [\sum_{i=1}^{L} r_i]} \alpha_k(\mathcal{I}_k) \det\left(\boldsymbol{\Phi}_{\mathcal{I}_k}\right), \tag{5.7}$$

where $\boldsymbol{\Phi}_{\mathcal{I}_k} \triangleq (\boldsymbol{\Sigma}_{\mathcal{I}_k}^{1/2})^{\text{H}}\mathbf{F}_{\text{RF}}\mathbf{F}_{\text{RF}}^{\text{H}}\boldsymbol{\Sigma}_{\mathcal{I}_k}^{1/2}$, and $\boldsymbol{\Sigma}_{\mathcal{I}_k}^{1/2}$ consists of columns selected by $\mathcal{I}_{i,k}$, $1 \le i \le L$ with $|\mathcal{I}_{i,k}| \le g_i$ from the ith block $\boldsymbol{\Sigma}_i^{1/2}$ of $\boldsymbol{\Sigma}^{1/2}$, and $\alpha_k(\mathcal{I}_k) = \prod_{i=1}^{L} \frac{g_i!}{(g_i - |\mathcal{I}_{i,k}|)!}$.

Proof See Appendix 3.

Since $\det(\cdot)$ is a smooth function, we may numerically optimize $\mathcal{J}(\mathbf{F}_{\text{RF}})$ using Newton-like algorithms. To ensure that an ascent direction is generated at each iteration, we adopt a trust-region framework. Remarkably, given the compactness of the manifold under consideration, such an improved Newton method achieves global convergence at a locally superlinear rate [14].

To generate a feasible sequence that converges to a critical point of $\mathcal{J}(\mathbf{F}_{\text{RF}})$, i.e., points where the Riemannian gradient vanishes, an update on a Grassmann manifold can be obtained from the νth iterate as

$$\mathbf{F}_{\text{RF}}^{\nu+1} = \mathcal{Q}\left(\mathbf{F}_{\text{RF}}^{\nu} + \boldsymbol{\eta}^{\nu}\right),$$

where $\mathcal{Q}(\cdot)$ extracts the unitary matrix from QR decomposition, and the tangent vector $\boldsymbol{\eta}^{\nu} \in \mathbb{C}^{N_{\mathrm{t}} \times n_{\mathrm{t}}}$ lies in $\mathcal{T}_{\mathbf{F}_{\mathrm{RF}}^{\nu}}\mathcal{M}_{\mathrm{gr}}$, the tangent space of $\mathbf{F}_{\mathrm{RF}}^{\nu}$. For Newton-like algorithms in the N-dimensional Euclidean space such as $\mathbb{R}^{N}$, the search direction is generally obtained by solving a quadratic approximation for the objective function at the current iterate. In the case of Riemannian manifolds, we similarly construct a quadratic approximation $\mathcal{J}_{\mathrm{quad}}^{\nu}(\cdot)$ within a small neighborhood of radius Δ^{ν} centered at the origin of $\mathcal{T}_{\mathbf{F}_{\mathrm{RF}}^{\nu}}\mathcal{M}_{\mathrm{gr}}$. Let grad $\mathcal{J}(\mathbf{F}_{\mathrm{RF}}^{\nu})$ and Hess $\mathcal{J}(\mathbf{F}_{\mathrm{RF}}^{\nu})[\boldsymbol{\eta}]$ be the Riemannian gradient and the Riemannian Hessian along the tangent vector $\boldsymbol{\eta}$, respectively. Then the quadratic approximation at $\mathbf{F}_{\mathrm{RF}}^{\nu}$ can be constructed as

$$\mathcal{J}_{\mathrm{quad}}^{\nu}(\boldsymbol{\eta}) = \mathcal{J}(\mathbf{F}_{\mathrm{RF}}^{\nu}) + \Re\mathrm{tr}\left(\boldsymbol{\eta}^{\mathrm{H}}\mathrm{grad}\,\mathcal{J}(\mathbf{F}_{\mathrm{RF}}^{\nu})\right) + \frac{1}{2}\Re\mathrm{tr}\left(\boldsymbol{\eta}^{\mathrm{H}}\mathrm{Hess}\,\mathcal{J}(\mathbf{F}_{\mathrm{RF}}^{\nu})\,[\boldsymbol{\eta}]\right), \tag{5.8}$$

which allows us to derive a search direction $\boldsymbol{\eta}^{\nu}$ by solving the following trust-region subproblem, i.e.,

$$\begin{aligned} \boldsymbol{\eta}^{\nu} \triangleq \ & \arg\max_{\boldsymbol{\eta}\in\mathcal{T}_{\mathbf{F}_{\mathrm{RF}}^{\nu}}\mathcal{M}_{\mathrm{gr}}} \mathcal{J}_{\mathrm{quad}}^{\nu}(\boldsymbol{\eta}) \\ & \text{s.t.} \quad \|\boldsymbol{\eta}\|_{\mathrm{F}} \le \Delta^{\nu}. \end{aligned} \tag{5.9}$$

For the purpose of defining gradient and Hessian, we endow the complex Euclidean space $\mathbb{C}^{N_{\mathrm{t}} \times n_{\mathrm{t}}}$ with the inner product $\Re\mathrm{tr}(\mathbf{A}^{\mathrm{H}}\mathbf{B})$ for $\mathbf{A}, \mathbf{B} \in \mathbb{C}^{N_{\mathrm{t}} \times n_{\mathrm{t}}}$. As shown in appendix section "Joint Hybrid MMSE-VP Precoding", the Riemannian gradient reads as

$$\mathrm{grad}\,\mathcal{J}(\mathbf{F}_{\mathrm{RF}}) = 2\mathcal{P}_{\mathbf{F}_{\mathrm{RF}}}^{\mathrm{gr}} \sum_{k=1}^{K} \gamma^{k} \sum_{\mathcal{I}_k \subset [\sum_i r_i]} \alpha_k(\mathcal{I}_k) \det\left(\boldsymbol{\Phi}_{\mathcal{I}_k}\right) \boldsymbol{\Xi}_{\mathcal{I}_k} \mathbf{F}_{\mathrm{RF}}, \tag{5.10}$$

where $\mathcal{P}_{\mathbf{F}_{\mathrm{RF}}}^{\mathrm{gr}} \triangleq \mathbf{I}_{N_{\mathrm{t}}} - \mathbf{F}_{\mathrm{RF}}\mathbf{F}_{\mathrm{RF}}^{\mathrm{H}}$ defines the orthogonal projection onto the tangent space $\mathcal{T}_{\mathbf{F}_{\mathrm{RF}}}\mathcal{M}_{\mathrm{gr}}$, and $\boldsymbol{\Xi}_{\mathcal{I}_k} \triangleq \boldsymbol{\Sigma}_{\mathcal{I}_k}^{1/2}\boldsymbol{\Phi}_{\mathcal{I}_k}^{-1}(\boldsymbol{\Sigma}_{\mathcal{I}_k}^{1/2})^{\mathrm{H}}$. Furthermore, the Riemannian Hessian along the tangent vector $\boldsymbol{\eta}$ is obtained as

$$\begin{aligned} \mathrm{Hess}\,\mathcal{J}(\mathbf{F}_{\mathrm{RF}})[\boldsymbol{\eta}] = {} & 2\sum_{k=1}^{K} \gamma^{k} \sum_{\mathcal{I}_k \subset [\sum_i r_i]} \alpha_k(\mathcal{I}_k) \det\left(\boldsymbol{\Phi}_{\mathcal{I}_k}\right) \\ & \times \mathcal{P}_{\mathbf{F}_{\mathrm{RF}}}^{\mathrm{gr}}\Big\{-\boldsymbol{\Xi}_{\mathcal{I}_k}\left(\boldsymbol{\eta}\mathbf{F}_{\mathrm{RF}}^{\mathrm{H}} + \mathbf{F}_{\mathrm{RF}}\boldsymbol{\eta}^{\mathrm{H}}\right)\boldsymbol{\Xi}_{\mathcal{I}_k}\mathbf{F}_{\mathrm{RF}} - \boldsymbol{\eta}\mathbf{F}_{\mathrm{RF}}^{\mathrm{H}}\boldsymbol{\Xi}_{\mathcal{I}_k}\mathbf{F}_{\mathrm{RF}} \\ & + \mathrm{tr}\left[\boldsymbol{\Xi}_{\mathcal{I}_k}\left(\boldsymbol{\eta}\mathbf{F}_{\mathrm{RF}}^{\mathrm{H}} + \mathbf{F}_{\mathrm{RF}}\boldsymbol{\eta}^{\mathrm{H}}\right)\right]\boldsymbol{\Xi}_{\mathcal{I}_k}\mathbf{F}_{\mathrm{RF}} + \boldsymbol{\Xi}_{\mathcal{I}_k}\boldsymbol{\eta}\Big\}. \end{aligned} \tag{5.11}$$

We note that the tangent space $\mathcal{T}_{\mathbf{F}_{\mathrm{RF}}^{\nu}}\mathcal{M}_{\mathrm{gr}}$ is isomorphic to the Euclidean space $\mathbb{C}^{N_t \times n_t}$ [14]. In other words, algorithms developed to solve the trust-region subproblem in the Euclidean space are expected to be applicable to (5.9). Because the Riemannian Hessian $\mathrm{Hess}\,\mathcal{J}(\mathbf{F}_{\mathrm{RF}}^{\nu})$ is defined through its operation on a tangent vector $\boldsymbol{\eta} \in \mathcal{T}_{\mathbf{F}_{\mathrm{RF}}^{\nu}}\mathcal{M}_{\mathrm{gr}}$, i.e., $\mathrm{Hess}\,\mathcal{J}(\mathbf{F}_{\mathrm{RF}}^{\nu})[\boldsymbol{\eta}]$, it cannot be directly formed on the tangent space $\mathcal{T}_{\mathbf{F}_{\mathrm{RF}}^{\nu}}\mathcal{M}_{\mathrm{gr}}$. Fortunately, when matrix-free techniques are considered which deal with matrix-vector products instead of direct matrix inversion or factorization [14, 15], only the knowledge of $\mathrm{Hess}\,\mathcal{J}(\mathbf{F}_{\mathrm{RF}}^{\nu})[\boldsymbol{\eta}]$ is needed. Furthermore, iteration of such techniques can be terminated as soon as an improved solution over the Cauchy point is found, which suffices to guarantee the superlinear convergence rate of the trust-region method [14]. In Algorithm 5.1, we adapt the Steihaug-Toint truncated conjugate-gradient (tCG) method from [14, 15] to our case to approximately solve (5.9). Such a step is integral to locating a critical point of $\mathcal{J}(\mathbf{F}_{\mathrm{RF}})$ by the trust-region Newton method, as formally stated in Algorithm 5.2.

Algorithm 5.1 Approximate solution to the trust-region subproblem via tCG method

Output: A search direction $\boldsymbol{\eta}^{\nu}$ at the current iterate $\mathbf{F}_{\mathrm{RF}}^{\nu}$.

1: Initialize $\boldsymbol{\eta}_0 \leftarrow 0$, $\boldsymbol{\mu}_0 \leftarrow \mathrm{grad}\,\mathcal{J}(\mathbf{F}_{\mathrm{RF}}^{\nu})$, $\boldsymbol{\delta}_0 \leftarrow -\boldsymbol{\mu}_0$, , $j \leftarrow 0$.
2: **while** true **do**
3: If $\Re\{\mathrm{tr}(\boldsymbol{\delta}_0^{\mathrm{H}}\mathrm{Hess}\,\mathcal{J}(\mathbf{F}_{\mathrm{RF}}^{\nu})[\boldsymbol{\delta}_0])\} \le 0$, compute $\tau \ge 0$ such that $\boldsymbol{\eta} \leftarrow \boldsymbol{\eta}_j + \tau\boldsymbol{\delta}_j$ maximizes $\mathcal{J}_{\mathrm{quad}}^{\nu}(\boldsymbol{\eta})$ in (5.8) subject to $||\boldsymbol{\eta}||_{\mathrm{F}} = \Delta^{\nu}$. Return $\boldsymbol{\eta}^{\nu} \leftarrow \boldsymbol{\eta}$;
4: Set $\alpha_j = \frac{\|\boldsymbol{\mu}_j\|_{\mathrm{F}}^2}{\Re\{\mathrm{tr}(\boldsymbol{\delta}_j^{\mathrm{H}}\mathrm{Hess}\,\mathcal{J}(\mathbf{F}_{\mathrm{RF}}^{\nu})[\boldsymbol{\delta}_j])\}}$ and $\boldsymbol{\eta}_{j+1} \leftarrow \boldsymbol{\eta}_j + \alpha_j\boldsymbol{\delta}_j$;
5: If $||\boldsymbol{\eta}_{j+1}||_{\mathrm{F}} \ge \Delta^{\nu}$, compute $\tau \ge 0$ such that $\boldsymbol{\eta} \leftarrow \boldsymbol{\eta}_j + \tau\boldsymbol{\delta}_j$ satisfies $||\boldsymbol{\eta}||_{\mathrm{F}} = \Delta^{\nu}$. Return $\boldsymbol{\eta}^{\nu} \leftarrow \boldsymbol{\eta}$;
6: Set $\boldsymbol{\mu}_{j+1} \leftarrow \boldsymbol{\mu}_j + \alpha_j\,\mathrm{Hess}\,\mathcal{J}\,(\mathbf{F}_{\mathrm{RF}})\,[\boldsymbol{\delta}_j]$;
7: If $\|\boldsymbol{\mu}_{j+1}\|_{\mathrm{F}} \le \|\boldsymbol{\mu}_0\|_{\mathrm{F}}\min(\|\boldsymbol{\mu}_0\|_{\mathrm{F}}^{1/2}, 0.1)$, return $\boldsymbol{\eta}^{\nu} \leftarrow \boldsymbol{\eta}_{j+1}$;
8: Set $\beta_{j+1} \leftarrow \frac{\|\boldsymbol{\mu}_{j+1}\|_{\mathrm{F}}^2}{\|\boldsymbol{\mu}_j\|_{\mathrm{F}}^2}$ and $\boldsymbol{\delta}_{j+1} \leftarrow -\boldsymbol{\mu}_{j+1} + \beta_{j+1}\boldsymbol{\delta}_j$;
9: Update $j \leftarrow j + 1$;
10: **end while**

Remark 5.3 In steps 3 and 5, computation of the parameter $\tau \ge 0$ can be recast as finding the positive root of the quadratic equation $\tau^2\|\boldsymbol{\delta}_j\|_{\mathrm{F}}^2 + 2\tau\Re\mathrm{tr}(\boldsymbol{\eta}_j^{\mathrm{H}}\boldsymbol{\delta}_j) = (\Delta^{\nu})^2 - \|\boldsymbol{\eta}_j\|_{\mathrm{F}}^2$ [14].

Remark 5.4 The global convergence of Algorithm 5.2 can be straightforwardly established following Corollary 7.4.6 in [14], which states that if the cost function is smooth and the Riemannian manifold is compact, trust-region Newton methods achieve global convergence. In our case, the conditions can be seen to be satisfied since the objective function, consisting of a weighted sum of $\det(\cdot)$, is smooth, and the Grassmann manifold $\mathcal{M}_{\mathrm{gr}}$ under consideration is compact.

Algorithm 5.2 Statistical CSI-based RF precoding via Riemannian manifold optimization

Output: A critical point $\mathbf{F}_{\mathrm{RF}}^{\nu}$ at which grad $\mathcal{J}(\mathbf{F}_{\mathrm{RF}}^{\nu}) < \epsilon$.

Initialization: $\mathbf{F}_{\mathrm{RF}}^{0}$, $\nu \leftarrow 0$, $\bar{\Delta} \leftarrow \sqrt{n_{\mathrm{t}}}$, $\Delta_0 \leftarrow 0.125\bar{\Delta}$, $\rho' \leftarrow 0.1$, tolerance ϵ.
while grad $\mathcal{J}(\mathbf{F}_{\mathrm{RF}}^{\nu}) < \epsilon$ **do**
    Obtain $\boldsymbol{\eta}^{\nu}$ by approximately solving (5.9) via Algorithm 5.1;
    Set $\rho^{\nu} \leftarrow \frac{\mathcal{J}(\mathbf{F}_{\mathrm{RF}}^{\nu}) - \mathcal{J}(\mathcal{Q}(\mathbf{F}_{\mathrm{RF}}^{\nu} + \boldsymbol{\eta}^{\nu}))}{\mathcal{J}_{\mathrm{quad}}^{\nu}(\mathbf{0}) - \mathcal{J}_{\mathrm{quad}}^{\nu}(\boldsymbol{\eta}^{\nu})}$;
    If $\rho^{\nu} \leq 0.25$, set $\Delta^{\nu+1} \leftarrow 0.25\Delta^{\nu}$;
    If $\rho^{\nu} \geq 0.75$ and $\|\boldsymbol{\eta}^{\nu}\|_{\mathrm{F}} = \Delta^{\nu}$, set $\Delta^{\nu+1} \leftarrow \min\{2\Delta^{\nu}, \bar{\Delta}\}$; otherwise set $\Delta^{\nu+1} \leftarrow \Delta^{\nu}$;
    If $\rho^{\nu} \geq \rho'$, set $\mathbf{F}_{\mathrm{RF}}^{\nu+1} \leftarrow \mathcal{Q}(\mathbf{F}_{\mathrm{RF}}^{\nu} + \boldsymbol{\eta}^{\nu})$; otherwise set $\mathbf{F}_{\mathrm{RF}}^{\nu+1} \leftarrow \mathbf{F}_{\mathrm{RF}}^{\nu}$;
    Update $\nu \leftarrow \nu + 1$;
end while

5.3.3 *Statistical CSI-Based RF Phase-Shifting Design*

It is sometimes desirable to employ only phase-shifting elements in the RF domain, which serves as a viable option for reducing the complexity of hardware implementation. It has been shown that the eigen-space of a large-dimensional Toeplitz matrix can be well approximated by the column space of a DFT matrix [9]. This implies that the L transmit correlation matrices $\{\boldsymbol{\Sigma}_i\}$ can be represented by a single DFT matrix. Although suboptimal, DFT beam selection simplifies beamforming network implementation and precoder update through limited feedback. Therefore, we consider DFT beamforming as an alternative to optimal constant-modulus RF design. Such simplification has also been exploited in [9, 16]. In this case, the RF precoding matrix is related to the DFT matrix by

$$\mathbf{F}_{\mathrm{RF}}\mathbf{F}_{\mathrm{RF}}^{\mathrm{H}} = \mathbf{T}\mathbf{Z}\mathbf{T}^{\mathrm{H}}, \tag{5.12}$$

where $\mathbf{T} = [\mathbf{t}_0, \ldots, \mathbf{t}_{N_{\mathrm{t}}-1}]$ denotes the N_{t}-dimensional DFT matrix with $[\mathbf{t}_k]_l \triangleq e^{-\jmath 2\pi kl/N_{\mathrm{t}}}, 0 \leq k, l \leq N_{\mathrm{t}} - 1$, and $\mathbf{Z} \triangleq \mathrm{diag}\{\mathbf{z}\}$ with $\mathbf{z} \triangleq [z_1, \ldots, z_{N_{\mathrm{t}}}]^{\mathrm{T}} \in \mathbb{B}^{N_{\mathrm{t}}}$. For user separation in the beam domain, it is required that $K \leq \sum_i z_i \leq n_{\mathrm{t}}$. Substituting (5.12) into (5.7), the problem for DFT codebook-based RF precoding is thus formulated as

$$\begin{aligned} \max_{\mathbf{z} \in \mathbb{B}^{N_{\mathrm{t}}}} \ & \mathcal{J}_{\mathrm{dft}}(\mathbf{z}) \triangleq \sum_{k=0}^{K} \gamma^k \sum_{\mathcal{I}_k = \bigcup_{i=1}^{L} \mathcal{I}_{i,k} \subset [\sum_i r_i]} \alpha_k(\mathcal{I}_k) \det\left(\boldsymbol{\Phi}_{\mathcal{I}_k}(\mathbf{z})\right) \\ \text{s.t.} \ & K \leq \sum_{i=1}^{N_{\mathrm{t}}} z_i \leq n_{\mathrm{t}}, \end{aligned} \tag{5.13}$$

where $\boldsymbol{\Phi}_{\mathcal{I}_k}(\mathbf{z}) \triangleq (\boldsymbol{\Sigma}_{\mathcal{I}_k}^{1/2})^{\mathrm{H}}\mathbf{TZT}^{\mathrm{H}}\boldsymbol{\Sigma}_{\mathcal{I}_k}^{1/2}$. By determinant expansion, it can be seen that the objective function is multi-linear in $\{z_1, \ldots, z_{N_t}\}$, and hence non-convex. Actually, such problems are NP-hard in the presence of the Boolean constraint. In view of the large number of antennas (on the order of tens and hundreds), a brute-force exhaustive search would not be computationally affordable. Fortunately, it turns out unnecessary to examine the entire feasible set by exploiting the problem structure.
Define $\mathcal{Z} \triangleq \{\mathbf{z} \in \mathbb{R}_+^{N_t} \mid \sum_i z_i - n_t \leq 0\}$, and note that $\mathcal{Z}$ is a compact normal set. To begin with, we make the following observation about the monotonicity of the objective function.

Proposition 5.1 *The objective function $\mathcal{J}_{\mathrm{dft}}(\mathbf{z})$ in (5.13) is smooth and increasing on $\mathcal{Z}$. Accordingly, the feasible set described by $\{\mathbf{z} \in \mathbb{B}^{N_t} \mid K \leq \sum_i z_i \leq n_t\}$ can be rewritten as $\mathcal{Z} \cap \mathbb{B}^{N_t}$ without loss of optimality.*

Proof See Appendix 5.

A search over $\mathcal{Z}\cap\mathbb{B}^{N_t}$ can be divided into searches over multiple subsets. Because $\mathcal{J}_{\mathrm{dft}}(\mathbf{z})$ is increasing and $\mathcal{Z}$ is compact, it suffices to test the lower and upper bounds of each subset to determine if an optimal solution is present. Further reduction can be performed on each candidate subset, where we seek a feasible solution to refine the current best value. This is the basic idea of BRnB [17].

Branching
Suppose that the current best value and the upper bound for $\mathcal{J}_{\mathrm{dft}}(\cdot)$ are $\mathcal{J}_{\mathrm{dft}}^{\mathrm{cbv}}$ and $\mathcal{J}_{\mathrm{dft}}^{\mathrm{ub}}$, respectively. Let $\mu(M)$ denote a local upper bound for the feasible solutions over $M = [\mathbf{p}, \mathbf{q}]$, and $\mathcal{A}_{[\mathbf{p},\mathbf{q}]}$ the set of positions at which $\mathbf{p}$ and $\mathbf{q}$ differ, i.e., $\mathcal{A}_{[\mathbf{p},\mathbf{q}]} \triangleq \{i \mid [\mathbf{p}]_i = 0, [\mathbf{q}]_i = 1, 1 \leq i \leq N_t\}$. The branching procedure starts with selecting from a set of box candidates $\mathcal{P}$ the one with the largest local upper bound, i.e., $M = \arg\max_{M' \in \mathcal{P}} \mu(M')$. Then M is partitioned at $j \in \mathcal{A}_{[\mathbf{p},\mathbf{q}]}$ into two smaller boxes as $M_0 = [\mathbf{p}_0, \mathbf{q}_0]$, $M_1 = [\mathbf{p}_1, \mathbf{q}_1]$, where $\mathbf{p}_m = \mathbf{p}|_{[\mathbf{p}]_j = m}$ and $\mathbf{q}_m = \mathbf{q}|_{[\mathbf{q}]_j = m}$ for $m = 0, 1$. Clearly, $M_0 \cap M_1 = \emptyset$, and $M = M_0 \cup M_1$. It was observed in [9] that the relevant DFT beams for, say $\mathcal{G}_i$, are related to its mean AoD $\bar{\theta}_i$ and angle spread Δ_i as

$$\mathcal{I}_{\mathrm{dft}}^i = \big\{m : -0.5\sin\big(\bar{\theta}_i + \Delta_i\big) \leq m/N_t - 0.5 \leq -0.5\sin\big(\bar{\theta}_i - \Delta_i\big), m = 0, \ldots, N_t - 1\big\}, \tag{5.14}$$

under the assumption that $-\frac{\pi}{2} \leq \bar{\theta}_i - \Delta_i < \bar{\theta}_i + \Delta_i \leq \frac{\pi}{2}$. Therefore, we may restrict the effective search space to be $\cup_{i=1}^{L} \mathcal{I}_{\mathrm{dft}}^i$ while giving priority to the non-overlapping DFT beams.

Reduction Let $h_{\mathrm{dft}}(\mathbf{z}) \triangleq \sum_i z_i - n_t$. In view of an increasing $h_{\mathrm{dft}}(\cdot)$ and $\mathcal{J}_{\mathrm{dft}}(\cdot)$, we infer that a box, say M_0, might contain feasible solutions if $h_{\mathrm{dft}}(\mathbf{p}_0) \leq 0$. Furthermore, it is only possible for such solutions to improve the current best one if the local upper bound satisfies $\mu(M_0) \geq \mathcal{J}_{\text{dft}}^{\text{cbv}}$. In this case, the search space can be reduced as

$$[\bar{\mathbf{p}}_0]_i \triangleq \begin{cases} 1, & \mathcal{J}_{\text{dft}}(\mathbf{q}_0 - \mathbf{e}_i) < \mathcal{J}_{\text{dft}}^{\text{cbv}} \\ 0, & \mathcal{J}_{\text{dft}}(\mathbf{q}_0 - \mathbf{e}_i) \geq \mathcal{J}_{\text{dft}}^{\text{cbv}} \end{cases}, \quad [\bar{\mathbf{q}}_0]_i \triangleq \begin{cases} 0, & h_{\text{dft}}(\bar{\mathbf{p}}_0 + \mathbf{e}_i) > 0 \\ 1, & h_{\text{dft}}(\bar{\mathbf{p}}_0 + \mathbf{e}_i) \leq 0 \end{cases}, \tag{5.15}$$

where the modification is performed successively, i.e., first on $\mathbf{p}_0$ at $i \in \mathcal{A}_{[\mathbf{p}_0, \mathbf{q}_0]}$ and then on $\mathbf{q}_0$ at $i \in \mathcal{A}_{[\bar{\mathbf{p}}_0, \mathbf{q}_0]}$. It can be seen that $\bar{M}_0 \triangleq [\bar{\mathbf{p}}_0, \bar{\mathbf{q}}_0] \subset [\mathbf{p}_0, \mathbf{q}_0]$. The following lemma establishes that there is no loss of optimality by such reduction.

Lemma 5.2 *If a feasible solution not worse than the current best solution is contained in M_0, then it must also be in $\bar{M}_0$.*

Proof See Appendix 6.

Bounding

To speed up convergence, computationally simple but tight global lower and upper bounds are desirable. Clearly, any feasible solution in $\bar{M}_0$ serves as a local lower bound $\eta(\bar{M}_0)$ while $\mathcal{J}_{\text{dft}}(\bar{\mathbf{q}}_0)$ can be used as a local upper bound due to monotonicity. In hopes of arriving at a good $\eta(\bar{M}_0)$, we note that the objective can be upper-bounded by

$$\mathcal{J}_{\text{dft}}(\mathbf{z}) = \mathbb{E}_{\{\mathbf{H}_{i,w}\}}\left[\det\left(\mathbf{I}_K + \gamma \mathbf{H}^{\text{H}}\mathbf{T}\mathbf{Z}\mathbf{T}^{\text{H}}\mathbf{H}\right)\right] \leq \underbrace{\left[1 + \frac{\gamma}{K}\text{tr}\left(\mathbf{T}^{\text{H}}\boldsymbol{\Sigma}_{\text{sum}}\mathbf{T}\mathbf{Z}\right)\right]^K}_{\triangleq \tilde{\mathcal{J}}_{\text{dft}}(\mathbf{z})}, \tag{5.16}$$

where $\boldsymbol{\Sigma}_{\text{sum}} \triangleq \sum_{i=1}^{L} g_i \boldsymbol{\Sigma}_i$, and the inequality $\det(\mathbf{A})^{\frac{1}{K}} \leq \frac{1}{K}\text{tr}(\mathbf{A})$ is applied. In doing this, a feasible solution $\mathbf{z}_0' \in \bar{M}_0$ maximizing $\tilde{\mathcal{J}}_{\text{dft}}(\cdot)$ can be easily obtained. Accordingly, a local upper bound is given as $\mu(\bar{M}_0) = \min\{\tilde{\mathcal{J}}_{\text{dft}}(\mathbf{z}_0'), \mathcal{J}_{\text{dft}}(\bar{\mathbf{q}}_0)\}$. The process of branching, reduction, and bounding is iterated until the gap between the current best value and upper bounds vanishes. This is formally stated in Algorithm 5.3.

Remark 5.5 The BRnB algorithm is bound to converge to a global optimal solution [17]. On one hand, because of the combinatorial nature of the problem, the algorithm will terminate after a finite number of iterations. On the other hand, because the current best solution, which is obtained by computing a local lower bound, improves after each iteration (as guaranteed by Step 9) and no optimality is lost from the steps of branching and reduction (as proven in Lemma 5.2), a global optimum will eventually be reached.

Algorithm 5.3 DFT codebook-based RF phase-shifting with JCP via BRnB

Output: A globally optimal solution $\mathbf{z}^{\star} \leftarrow \mathbf{z}^{\text{cbv}}$.
1: Initialization: $\mathcal{P} \leftarrow \{[\mathbf{0}, \mathbf{1}]\}$, $\mathcal{J}_{\text{dft}}^{\text{cbv}} \leftarrow 0$, $\mathcal{J}_{\text{dft}}^{\text{ub}} \leftarrow \infty$.
2: **while** $\mathcal{J}_{\text{dft}}^{\text{ub}} > \mathcal{J}_{\text{dft}}^{\text{cbv}}$ **do**
3: Delete the infeasible boxes in $\mathcal{P}$;
4: Select $M = [\mathbf{p}, \mathbf{q}] \in \mathcal{P}$ with the largest local upper bound $M = \arg\max_{M' \in \mathcal{P}} \mu(M')$;
5: Branch M into M_0, M_1, and update $\mathcal{P} \leftarrow \mathcal{P} \backslash M$;
6: **for** $i = 0, 1$ **do**
7: Reduce $M_i = [\mathbf{p}_i, \mathbf{q}_i]$ to $\bar{M}_i = [\bar{\mathbf{p}}_i, \bar{\mathbf{q}}_i]$ according to (5.15);
8: Compute a local solution $\mathbf{z}_i' \in \bar{M}_i$ by maximizing (5.16), a local lower bound $\eta(\bar{M}_i) \leftarrow \mathcal{J}_{\text{dft}}(\mathbf{z}_i')$, and a local upper bound $\mu(\bar{M}_i) \leftarrow \min\{\tilde{\mathcal{J}}_{\text{dft}}(\mathbf{z}_0'), \mathcal{J}_{\text{dft}}(\bar{\mathbf{q}}_i)\}$;
9: Update $\mathcal{P} \leftarrow \mathcal{P} \cup \bar{M}_i$;
10: If $\eta(\bar{M}_i) > \mathcal{J}_{\text{dft}}^{\text{cbv}}$, update $\mathcal{J}_{\text{dft}}^{\text{cbv}} \leftarrow \eta(\bar{M}_i)$ and $\mathbf{z}^{\text{cbv}} \leftarrow \mathbf{z}_i'$;
11: **end for**
12: Update the global upper bound $\mathcal{J}_{\text{dft}}^{\text{ub}} \leftarrow \max_{M' \in \mathcal{P}} \mu(M')$;
13: **end while**

5.4 Cluster-Wise Hybrid MMSE-VP Precoding

Joint baseband processing, though delivering superior performance gain, requires knowledge of the direct effective channel $\mathbf{H}_i\mathbf{F}_{\text{RF},i}$ for self-transmission optimization as well as the cross effective channels $\mathbf{H}_j\mathbf{F}_{\text{RF},i}$, $\forall j \neq i$, for interference management. To alleviate the overhead of channel training, one possibility is to focus baseband precoding on each cluster as long as the RF stage effectively separates multiple clusters in the beam domain. Specifically, RF precoding should aim for a trade-off between self-transmission and interference leakage control. In this section, we first formulate the RF design problem upon deriving the baseband solutions for PCP. We further show that the nonuniform- and constant-modulus cases can be solved by adapting the solution techniques developed in Sect. 5.3.

5.4.1 MSE-Based Problem Formulation

Mathematically, the aggregate baseband precoder for PCP is block diagonal, i.e., $\mathbf{F}_{\text{BB}} = \text{diag}\{\mathbf{F}_{\text{BB},1}, \ldots, \mathbf{F}_{\text{BB},L}\}$ with $\mathbf{F}_{\text{BB},i}$, $1 \leq i \leq L$, as the baseband precoder for the ith cluster. Recall the conditional cluster-wise MSE $\mathcal{E}_i[t]$ in (5.1) as

$$\begin{aligned}\mathcal{E}_i[t] &= \mathbb{E}_{\mathbf{n}_i[t]}\left[\left\|\beta^{-1}\left(\mathbf{H}_i^{\text{H}}\mathbf{F}_{\text{RF},i}\mathbf{x}_i[t] + \mathbf{n}_i[t]\right) - \mathbf{s}_i[t]\right\|^2 \middle| \mathbf{d}_i[t]\right] \\ &= \beta^{-2}\left(\mathbf{x}_i[t]^{\text{H}}\mathbf{F}_{\text{RF},i}^{\text{H}}\mathbf{H}_i\mathbf{H}_i^{\text{H}}\mathbf{F}_{\text{RF},i}\mathbf{x}_i[t] + g_i\sigma_n^2\right)\end{aligned}$$

$$- 2\beta^{-1}\Re\left\{\mathbf{s}_i[t]\mathbf{H}_i^{\mathrm{H}}\mathbf{F}_{\mathrm{RF},i}\mathbf{x}_i[t]\right\} + \|\mathbf{s}_i[t]\|^2.$$

The problem of hybrid design in (5.2) is reformulated as

$$\begin{aligned}
&\min_{\mathbf{F}_{\mathrm{RF}}} \mathbb{E}_{\{\mathbf{H}_{i,w},\mathbf{d}_i[t]\}}\left[\min_{\{\mathbf{x}_i[t],\mathbf{q}_i[t]\}} \frac{1}{T}\sum_{i=1}^{L}\sum_{t\in\mathcal{T}_n}\mathcal{E}_i[t]\right] \\
&\text{s.t. } \mathbf{F}_{\mathrm{RF}}^{\mathrm{H}}\mathbf{F}_{\mathrm{RF}} = \mathbf{I}_{n_t},\ \frac{1}{T}\sum_{i=1}^{L}\sum_{t\in\mathcal{T}_n}\|\mathbf{x}_i[t]\|^2 \le P_{\mathrm{t}}. \qquad (5.17)
\end{aligned}$$

Following a similar line of derivation as in [10], solving the KKT conditions leads to

$$\mathbf{x}_i^{\star}[t] = \beta^{\star}\mathbf{F}_{\mathrm{BB},i}^{\star}\mathbf{s}_i[t],$$

where $\mathbf{F}_{\mathrm{BB},i}^{\star}$ is the cluster-wise MMSE baseband precoder, i.e,

$$\mathbf{F}_{\mathrm{BB},i}^{\star} = \left(\mathbf{F}_{\mathrm{RF},i}^{\mathrm{H}}\mathbf{H}_i\mathbf{H}_i^{\mathrm{H}}\mathbf{F}_{\mathrm{RF},i} + \gamma^{-1}\mathbf{I}_{g_i}\right)^{-1}\mathbf{F}_{\mathrm{RF},i}^{\mathrm{H}}\mathbf{H}_i,$$

and the total power constraint is enforced by setting the scaling factor

$$\beta^{\star} = \sqrt{\frac{P_{\mathrm{t}}}{\frac{1}{T}\sum_{i=1}^{L}\sum_{t\in\mathcal{T}_n}\left\|\mathbf{F}_{\mathrm{BB},i}^{\star}\mathbf{s}_i[t]\right\|^2}}.$$

Note that although baseband precoding is performed with respect to each cluster, the scaling factor β is nonetheless shared between all users. As a result, the MSE is simplified as

$$\begin{aligned}
\mathcal{E}_i[t] &= \mathbf{s}_i[t]^{\mathrm{H}}\left[\mathbf{I}_{g_i} - \mathbf{H}_i^{\mathrm{H}}\mathbf{F}_{\mathrm{RF},i}\left(\mathbf{F}_{\mathrm{RF},i}^{\mathrm{H}}\mathbf{H}_i\mathbf{H}_i^{\mathrm{H}}\mathbf{F}_{\mathrm{RF},i} + \gamma\mathbf{I}_{g_i}\right)^{-1}\mathbf{F}_{\mathrm{RF},i}^{\mathrm{H}}\mathbf{H}_i\right]\mathbf{s}_i[t] \\
&= \left(\mathbf{d}_i[t] + \tau\mathbf{q}_i[t]\right)^{\mathrm{H}}\mathbf{B}_i^{-1}\left(\mathbf{d}_i[t] + \tau\mathbf{q}_i[t]\right),
\end{aligned}$$

where $\mathbf{B}_i = \mathbf{I}_{g_i} + \gamma\mathbf{H}_i^{\mathrm{H}}\mathbf{F}_{\mathrm{RF},i}\mathbf{F}_{\mathrm{RF},i}^{\mathrm{H}}\mathbf{H}_i$, and the last step is obtained by the matrix inversion lemma. The perturbation vector is found by solving the following integer LS problem

$$\begin{aligned}
\mathbf{q}_i^{\star}[t] &= \operatorname*{arg\,min}_{\mathbf{q}_i[t]\in\mathbb{Z}^{g_i}+\jmath\mathbb{Z}^{g_i}} \mathbf{s}_i[t]^{\mathrm{H}}\mathbf{B}_i^{-1}\mathbf{s}_i[t] \\
&= \operatorname*{arg\,min}_{\mathbf{q}_i[t]\in\mathbb{Z}^{g_i}+\jmath\mathbb{Z}^{g_i}} \|-\mathbf{L}_i\mathbf{d}_i[t] - \tau\mathbf{L}_i\mathbf{q}_i[t]\|^2,
\end{aligned}$$

where the lower triangular $\mathbf{L}_i \in \mathbb{C}^{g_i \times g_i}$ is obtained by Cholesky decomposition $\mathbf{B}_i^{-1} = \mathbf{L}_i^{\mathrm{H}} \mathbf{L}_i$. Let $\mathcal{E}_i^{\star[t]}(\mathbf{F}_{\mathrm{RF},i})$ denote the resulting MSE. Thus, the problem (5.17) is reduced as

$$\begin{aligned} &\min_{\mathbf{F}_{\mathrm{RF}}} \frac{1}{T} \sum_{t \in \mathcal{T}_n} \sum_{i=1}^{L} \mathbb{E}_{\mathbf{H}_{i,w},\mathbf{d}_i[t]} \left[\mathcal{E}_i^{\star[t]} \left(\mathbf{F}_{\mathrm{RF},i} \right) \right] \\ &\text{s.t. } \mathbf{F}_{\mathrm{RF}}^{\mathrm{H}} \mathbf{F}_{\mathrm{RF}} = \mathbf{I}_{n_{\mathrm{t}}}. \end{aligned} \tag{5.18}$$

5.4.2 Modified MSE-Based RF Precoding Design

The feasibility of deriving the baseband solutions without regard to inter-cluster interference hinges on the effectiveness of the RF design to separate different user clusters in the beam domain. Although solving (5.18) gives one possible solution, such selfish optimization could be problematic when there is a significant overlap between the channel correlated subspaces. Instead, it makes more sense for RF precoder design to strike a balance between optimizing self-transmission and the interference caused to others. Mathematically, the former can be represented by the objective function of the average MSE $\mathbb{E}_{\mathbf{H}_{i,w},\mathbf{d}_i[t]}[\mathcal{E}_i^{\star[t]}(\mathbf{F}_{\mathrm{RF},i})]$ and latter by the leakage power Ω_i, which is defined as

$$\begin{aligned} \Omega_i &\triangleq \sum_{j=1, j \neq i}^{L} \mathbb{E}_{\mathbf{H}_{j,w}} \left[\left\| \mathbf{H}_j^{\mathrm{H}} \mathbf{F}_{\mathrm{RF},i} \right\|_{\mathrm{F}}^2 \right] \\ &= \sum_{j=1, j \neq i}^{L} g_j \operatorname{tr} \left(\mathbf{F}_{\mathrm{RF},i}^{\mathrm{H}} \boldsymbol{\Sigma}_j \mathbf{F}_{\mathrm{RF},i} \right). \end{aligned}$$

Accordingly, the trade-off for the ith cluster translates to minimization of

$$\mathcal{E}_{\mathrm{pcp},i} \triangleq \frac{1}{T} \sum_{t \in \mathcal{T}_n} \mathbb{E}_{\mathbf{H}_{i,w},\mathbf{d}_i[t]} \left[\mathcal{E}_i^{\star[t]}(\mathbf{F}_{\mathrm{RF},i}) \right] + \Omega_i.$$

Overall, joint RF precoding across all clusters is posed as a solution to the following optimization problem:

$$\begin{aligned} &\min_{\mathbf{F}_{\mathrm{RF}}} \mathcal{E}_{\mathrm{pcp}} \triangleq \sum_{i=1}^{L} \mathcal{E}_{\mathrm{pcp},i} = \frac{1}{T} \sum_{i=1}^{L} \sum_{t \in \mathcal{T}_n} \mathbb{E}_{\mathbf{H}_{i,w},\mathbf{d}_i[t]} \left[\mathcal{E}_i^{\star[t]}(\mathbf{F}_{\mathrm{RF},i}) \right] + \sum_{i=1}^{L} \Omega_i \\ &\text{s.t. } \mathbf{F}_{\mathrm{RF}}^{\mathrm{H}} \mathbf{F}_{\mathrm{RF}} = \mathbf{I}_{n_{\mathrm{t}}}. \end{aligned}$$

It is easy to see that the above formulation includes (5.2) as a special case. Note that different from the case of JCP, unitary transformation of an optimal solution is no longer optimal. In other words, the feasible set of $\mathbf{F}_{\mathrm{RF}}$ in fact constitutes a Stiefel manifold $\mathcal{M}_{\mathrm{st}} \triangleq \{\mathbf{F}_{\mathrm{RF}} \in \mathbb{C}^{N_{\mathrm{t}} \times n_{\mathrm{t}}} \,|\, \mathbf{F}_{\mathrm{RF}}^{\mathrm{H}} \mathbf{F}_{\mathrm{RF}} = \mathbf{I}_{n_{\mathrm{t}}}\}$. By evaluating the expectation to a closed form, the framework of trust-region optimization on smooth Riemannian manifold can also be exploited. Following a similar line of argument as in Theorem 5.2, a lower bound for the sum MSE can be derived as

$$\begin{aligned}
\frac{1}{T}\sum_{t\in\mathcal{T}_n}\sum_{i=1}^{L}\mathbb{E}\left[\mathcal{E}_i^{\star[t]}(\mathbf{F}_{\mathrm{RF},i})\right] &\geq \sum_{i=1}^{L}\frac{g_i\tau^2}{\pi e}\mathbb{E}\left\{\left[\det\left(\mathbf{I}_{g_i}+\gamma\mathbf{H}_i^{\mathrm{H}}\mathbf{F}_{\mathrm{RF},i}\mathbf{F}_{\mathrm{RF},i}^{\mathrm{H}}\mathbf{H}_i\right)\right]^{-\frac{1}{g_i}}\right\} \\
&\geq \sum_{i=1}^{L}\frac{g_i\tau^2}{\pi e}\left\{\mathbb{E}\left[\det\left(\mathbf{I}_{g_i}+\gamma\mathbf{H}_i^{\mathrm{H}}\mathbf{F}_{\mathrm{RF},i}\mathbf{F}_{\mathrm{RF},i}^{\mathrm{H}}\mathbf{H}_i\right)\right]\right\}^{-\frac{1}{g_i}} \\
&= \sum_{i=1}^{L}\frac{g_i\tau^2}{\pi e}\left[\sum_{k=0}^{g_i}\gamma^k\sum_{\mathcal{I}_k\subset[r_i]}\alpha_{i,k}\det\left(\boldsymbol{\Phi}_{i,\mathcal{I}_k}\right)\right]^{-\frac{1}{g_i}},
\end{aligned}$$

where $\boldsymbol{\Phi}_{i,\mathcal{I}_k} \triangleq (\boldsymbol{\Sigma}_{i,\mathcal{I}_k}^{1/2})^{\mathrm{H}}\mathbf{F}_{\mathrm{RF},i}\mathbf{F}_{\mathrm{RF},i}^{\mathrm{H}}\boldsymbol{\Sigma}_{i,\mathcal{I}_k}^{1/2}$ with $\boldsymbol{\Sigma}_{i,\mathcal{I}_k}^{1/2}$ consisting of columns $\mathcal{I}_k$ of $\boldsymbol{\Sigma}_i^{1/2}$, and $\alpha_{i,k} \triangleq \frac{g_i!}{(g_i-k)!}$. On the other hand, the total interference leakage power is expressed as

$$\begin{aligned}
\sum_{i=1}^{L}\Omega_i &= \sum_{i=1}^{L}\sum_{j=1,j\neq i}^{L} g_j\mathrm{tr}\left(\mathbf{F}_{\mathrm{RF},i}^{\mathrm{H}}\boldsymbol{\Sigma}_j\mathbf{F}_{\mathrm{RF},i}\right) \\
&= \sum_{i=1}^{L}\mathrm{tr}\left(\mathbf{F}_{\mathrm{RF},i}^{\mathrm{H}}\overline{\boldsymbol{\Sigma}}_i\mathbf{F}_{\mathrm{RF},i}\right),
\end{aligned}$$

where $\overline{\boldsymbol{\Sigma}}_i \triangleq \sum_{j=1,j\neq i}^{L} g_j\boldsymbol{\Sigma}_j$. For mathematical tractability, we study the lower bound in place of the original objective function, i.e.,

$$\begin{aligned}
\min_{\mathbf{F}_{\mathrm{RF}}\in\mathcal{M}_{\mathrm{st}}} \quad \mathcal{E}_{\mathrm{pcp,lb}}\left(\mathbf{F}_{\mathrm{RF}}\right) \triangleq \sum_{i=1}^{L}\frac{g_i\tau^2}{\pi e}\left[\sum_{k=0}^{g_i}\gamma^k\sum_{\mathcal{I}_k\subset[r_i]}\alpha_{i,k}\det\left(\boldsymbol{\Phi}_{i,\mathcal{I}_k}\right)\right]^{-1/g_i} \\
+\sum_{i=1}^{L}\mathrm{tr}\left(\mathbf{F}_{\mathrm{RF},i}^{\mathrm{H}}\overline{\boldsymbol{\Sigma}}_i\mathbf{F}_{\mathrm{RF},i}\right). \qquad (5.19)
\end{aligned}$$

Following a similar derivation as in the case of JCP, the Riemannian gradient and the Riemannian Hessian of the Stiefel manifold $\mathbf{F}_{\mathrm{RF}} \in \mathcal{M}_{\mathrm{st}}$ can be obtained from their

Euclidean counterparts. In particular, the derivation in appendix section "Cluster-Wise Hybrid MMSE-VP Precoding" yields the Riemannian gradient as

$$\begin{aligned}\operatorname{grad}\mathcal{E}_{\mathrm{pcp,lb}}\left(\mathbf{F}_{\mathrm{RF}}\right) &= \nabla\mathcal{E}_{\mathrm{pcp,lb}}\left(\mathbf{F}_{\mathrm{RF}}\right)\\ &\quad - \mathbf{F}_{\mathrm{RF}}\frac{1}{2}\left(\mathbf{F}_{\mathrm{RF}}^{\mathrm{H}}\nabla\mathcal{E}_{\mathrm{pcp,lb}}\left(\mathbf{F}_{\mathrm{RF}}\right) + \nabla^{\mathrm{H}}\mathcal{E}_{\mathrm{pcp,lb}}\left(\mathbf{F}_{\mathrm{RF}}\right)\mathbf{F}_{\mathrm{RF}}\right),\end{aligned} \tag{5.20}$$

where $\nabla\mathcal{E}_{\mathrm{pcp,lb}}\left(\mathbf{F}_{\mathrm{RF}}\right)$ is the Euclidean gradient of $\mathcal{E}_{\mathrm{pcp,lb}}\left(\mathbf{F}_{\mathrm{RF}}\right)$. In addition, the Riemannian Hessian is obtained as

$$\begin{aligned}\operatorname{Hess}\mathcal{E}_{\mathrm{pcp,lb}}\left(\mathbf{F}_{\mathrm{RF}}\right)[\boldsymbol{\eta}] &= \mathcal{P}_{\mathbf{F}_{\mathrm{RF}}}^{\mathrm{st}}\nabla^2\mathcal{E}_{\mathrm{pcp,lb}}\left(\mathbf{F}_{\mathrm{RF}}\right)[\boldsymbol{\eta}]\\ &\quad + \mathfrak{A}_{\mathbf{F}_{\mathrm{RF}}}^{\mathrm{st}}\left(\boldsymbol{\eta}, \nabla\mathcal{E}_{\mathrm{pcp,lb}}\left(\mathbf{F}_{\mathrm{RF}}\right) - \mathcal{P}_{\mathbf{F}_{\mathrm{RF}}}^{\mathrm{st}}\nabla\mathcal{E}_{\mathrm{pcp,lb}}\left(\mathbf{F}_{\mathrm{RF}}\right)\right)\end{aligned} \tag{5.21}$$

with $\nabla^2\mathcal{E}_{\mathrm{pcp,lb}}\left(\mathbf{F}_{\mathrm{RF}}\right)[\boldsymbol{\eta}]$ denoting the Euclidean Hessian along the direction $\boldsymbol{\eta}$, $\mathfrak{A}_{\mathbf{F}_{\mathrm{RF}}}^{\mathrm{st}}$ the Weingarten map [18], and $\mathcal{P}_{\mathbf{F}_{\mathrm{RF}}}^{\mathrm{st}}$ the orthogonal projector onto the tangent space $\mathcal{T}_{\mathbf{F}_{\mathrm{RF}}}\mathcal{M}_{\mathrm{st}}$. To particularize the trust-region framework in Algorithm 5.2 to the PCP case, we accordingly replace the objective function $\mathcal{J}(\mathbf{F}_{\mathrm{RF}})$ by $\mathcal{E}_{\mathrm{pcp,lb}}(\mathbf{F}_{\mathrm{RF}})$ in (5.19), the Riemannian gradient $\operatorname{grad}\mathcal{J}(\mathbf{F}_{\mathrm{RF}})$ by $\operatorname{grad}\mathcal{E}_{\mathrm{pcp,lb}}\left(\mathbf{F}_{\mathrm{RF}}\right)$ in (5.20) and the Riemannian Hessian $\operatorname{Hess}\mathcal{J}(\mathbf{F}_{\mathrm{RF}})[\boldsymbol{\eta}]$ by $\operatorname{Hess}\mathcal{E}_{\mathrm{pcp,lb}}\left(\mathbf{F}_{\mathrm{RF}}\right)[\boldsymbol{\eta}]$ in (5.21).

5.4.3 Statistical CSI-Based RF Phase-Shifting Design

In this section, we consider limited-feedback constant-modulus RF precoder design, which is based on a DFT codebook. In this case, the RF precoder for the ith cluster satisfies $\mathbf{F}_{\mathrm{RF},i}\mathbf{F}_{\mathrm{RF},i}^{\mathrm{H}} = \mathbf{T}\mathbf{Z}_i\mathbf{T}^{\mathrm{H}}$ where the nonzero diagonal elements of $\mathbf{Z}_i$ correspond to the chosen DFT beams. Let $\mathbf{z} \triangleq [\mathbf{z}_1^{\mathrm{T}}, \ldots, \mathbf{z}_L^{\mathrm{T}}]^{\mathrm{T}} \in \mathbb{B}^{LN_{\mathrm{t}}}$ with $\mathbf{z}_i \triangleq \operatorname{diag}(\mathbf{Z}_i)$, $1 \leq i \leq L$. We rewrite the objective function from (5.19) as

$$\mathcal{E}_{\mathrm{dft}}^{\mathrm{pcp}}(\mathbf{z}) \triangleq \mathcal{E}_{\mathrm{dft}}^{+}(\mathbf{z}) - \mathcal{E}_{\mathrm{dft}}^{-}(\mathbf{z}),$$

where $\mathcal{E}_{\mathrm{dft}}^{-}(\mathbf{z}) \triangleq \sum_{i=1}^{L}\operatorname{tr}(\mathbf{T}^{\mathrm{H}}\boldsymbol{\Sigma}_i\mathbf{T}\mathbf{Z}_i)$, and

$$\mathcal{E}_{\mathrm{dft}}^{+}(\mathbf{z}) \triangleq -\sum_{i=1}^{L}\frac{g_i\tau^2}{\pi e}\left[\sum_{k=0}^{g_i}\gamma^k\sum_{\mathcal{I}_k\subset[r_i]}\alpha_{i,k}\det\left((\boldsymbol{\Sigma}_{i,\mathcal{I}_k}^{1/2})^{\mathrm{H}}\mathbf{T}\mathbf{Z}_i\mathbf{T}^{\mathrm{H}}\boldsymbol{\Sigma}_{i,\mathcal{I}_k}^{1/2}\right)\right]^{-1/g_i}.$$

Accordingly, the problem is stated as

$$\max_{\mathbf{z}\in\mathbb{B}^{LN_t}} \mathcal{E}_{\text{dft}}^{\text{pcp}}(\mathbf{z})$$

$$\text{s.t.} \sum_{i=1}^{L} \mathbf{z}_i - \mathbf{1} \le \mathbf{0}, \sum_{i=1}^{L} \mathbf{1}^{\text{T}}\mathbf{z}_i - n_t \le 0, \mathbf{1}^{\text{T}}\mathbf{z}_i - g_i \ge 0, 1 \le i \le L, \quad (5.22)$$

where the cardinality constraints ensure user separation in the n_t-dimensional beam domain. Let $h_-(\mathbf{z}) \triangleq \max\{\{\sum_{i=1}^{L}[\mathbf{z}_i]_j - 1\}_{j=1}^{N_t}, \sum_{i=1}^{L}\mathbf{1}^{\text{T}}\mathbf{z}_i - n_t\}$ and $h_+(\mathbf{z}) \triangleq \min_{1\le i\le L}\{\mathbf{1}^{\text{T}}\mathbf{z}_i - g_i\}$. Clearly, (5.22) is equivalent to

$$\max_{\mathbf{z}\in\mathbb{B}^{LN_t}} \mathcal{E}_{\text{dft}}^{\text{pcp}}(\mathbf{z})$$

$$\text{s.t. } h_-(\mathbf{z}) \le 0 \le h_+(\mathbf{z}). \quad (5.23)$$

Note that both $h_-(\mathbf{z})$ and $h_+(\mathbf{z})$ are increasing functions as a result of the component constraint functions being linear with non-negative coefficients. The same can also be said about $\mathcal{E}_{\text{dft}}^{-}(\mathbf{z})$ since $\text{diag}(\mathbf{T}^{\text{H}}\mathbf{\Sigma}_i\mathbf{T}) \ge \mathbf{0}, i = 1, \ldots, L$. Following the line of reasoning of Proposition 5.1, $\mathcal{E}_{\text{dft}}^{+}(\mathbf{z})$ can be shown to be increasing. In summary, the problem in (5.23) has as an objective a difference of increasing (d.i.) functions and increasing functions as constraints. Interestingly, the technique of BRnB proposed in Sect. 5.3.3 is still applicable to solve such a problem to global optimality. Because the objective function is no longer monotonically increasing, the reduction and bounding stages need to be accordingly modified.

Branching
For ease of exposition, let us consider the matrix domain $\mathbb{B}^{N_t\times L}$ derived from reshaping the vector domain $\mathbb{B}^{LN_t}$ in a column-wise manner, i.e., for $\mathbf{x} \triangleq [\mathbf{x}_1^{\text{T}}, \ldots, \mathbf{x}_L^{\text{T}}]^{\text{T}} \in \mathbb{B}^{LN_t}$, we define $\text{vec}^{-1}(\mathbf{x}) = [\mathbf{x}_1, \ldots, \mathbf{x}_L] \in \mathbb{B}^{N_t\times L}$. Corresponding to $M = [\mathbf{p}, \mathbf{q}] \subset \mathbb{B}^{LN_t}$, row-wise branching is performed on $\tilde{M} \triangleq [\text{vec}^{-1}(\mathbf{p}), \text{vec}^{-1}(\mathbf{q})] \subset \mathbb{B}^{N_t\times L}$. Specifically, let $1 \le j \le N_t$ denote a row that has not been chosen for branching. We obtain $(L+1)$ smaller boxes $\tilde{M}_0, \ldots, \tilde{M}_L$ by replacing the jth row of $\text{vec}^{-1}(\mathbf{p})$ and $\text{vec}^{-1}(\mathbf{q})$ by $\mathbf{0}^{\text{T}}, \mathbf{e}_1^{\text{T}}, \ldots, \mathbf{e}_L^{\text{T}}$, respectively. In light of the constraint $\sum_{i=1}^{L}\mathbf{z}_i - \mathbf{1} \le \mathbf{0}$, it is not difficult to see that there is no loss of feasible solutions.

Reduction
Let $[\mathbf{p}, \mathbf{q}] \subset \mathbb{B}^{LN_t}$ be a box generated from the branching step. The necessary conditions for it to contain an optimal solution read as

$$h_-(\mathbf{p}) \le 0 \le h_+(\mathbf{q}), \; \mathcal{E}_{\text{dft}}^{+}(\mathbf{q}) - \mathcal{E}_{\text{dft}}^{-}(\mathbf{p}) \ge \mathcal{E}_{\text{dft}}^{\text{cbv}},$$

where $\mathcal{E}_{\text{dft}}^{\text{cbv}}$ denotes the current best value. The first condition tests if any feasible solution can be found in $[\mathbf{p}, \mathbf{q}]$, whose superiority over the current best solution is guaranteed by satisfying the second condition. If not, we have $\mathcal{E}_{\text{dft}}^{+}(\mathbf{x}) - \mathcal{E}_{\text{dft}}^{-}(\mathbf{x}) \le$

$\mathcal{E}^{+}_{\text{dft}}(\mathbf{q}) - \mathcal{E}^{-}_{\text{dft}}(\mathbf{p}) < \mathcal{E}^{\text{cbv}}_{\text{dft}}, \forall \mathbf{x} \in [\mathbf{p}, \mathbf{q}]$. If $[\mathbf{p}, \mathbf{q}]$ is worth further exploration, it is not necessary to explore it in its entirety, as shown in the following lemma.

Lemma 5.3 *If a feasible solution is contained in the box* $[\mathbf{p}, \mathbf{q}]$, *then it must also be contained in the reduced box* $[\bar{\mathbf{p}}, \bar{\mathbf{q}}] \subset [\mathbf{p}, \mathbf{q}]$, *where* $[\bar{\mathbf{p}}]_j, \forall j \in \{i \mid [\mathbf{p}]_i = 0, [\mathbf{q}]_i = 1, 1 \le i \le LN_{\text{t}}\}$ *is given as*

$$[\bar{\mathbf{p}}]_j \triangleq \begin{cases} 0, & h_{+}\left(\mathbf{q} - \mathbf{e}_j\right) \ge 0 \text{ and } \mathcal{E}^{+}_{\text{dft}}\left(\mathbf{q} - \mathbf{e}_j\right) - \mathcal{E}^{-}_{\text{dft}}\left(\mathbf{p}\right) \ge \mathcal{E}^{\text{cbv}}_{\text{dft}} \\ 1, & h_{+}\left(\mathbf{q} - \mathbf{e}_j\right) < 0 \text{ or } \mathcal{E}^{+}_{\text{dft}}\left(\mathbf{q} - \mathbf{e}_j\right) - \mathcal{E}^{-}_{\text{dft}}\left(\mathbf{p}\right) < \mathcal{E}^{\text{cbv}}_{\text{dft}} \end{cases}. \tag{5.24}$$

If $h_{-}(\bar{\mathbf{p}}) \le 0$, *we further modify* $[\mathbf{q}]_j, \forall j \in \{i \mid [\bar{\mathbf{p}}]_i = 0, [\mathbf{q}]_i = 1, 1 \le i \le LN_{\text{t}}\}$ *as*

$$[\bar{\mathbf{q}}]_j \triangleq \begin{cases} 1, & h_{-}\left(\bar{\mathbf{p}} + \mathbf{e}_j\right) \le 0 \text{ and } \mathcal{E}^{+}_{\text{dft}}\left(\mathbf{q}\right) - \mathcal{E}^{-}_{\text{dft}}\left(\bar{\mathbf{p}} + \mathbf{e}_j\right) \ge \mathcal{E}^{\text{cbv}}_{\text{dft}} \\ 0, & h_{-}\left(\bar{\mathbf{p}} + \mathbf{e}_j\right) > 0 \text{ or } \mathcal{E}^{+}_{\text{dft}}\left(\mathbf{q}\right) - \mathcal{E}^{-}_{\text{dft}}\left(\bar{\mathbf{p}} + \mathbf{e}_j\right) < \mathcal{E}^{\text{cbv}}_{\text{dft}} \end{cases}.$$

Proof See Appendix 6.

Bounding

Because of the complicating cardinality constraints, it is not straightforward to find a feasible solution within $M = [\mathbf{p}, \mathbf{q}]$ by greedy heuristics. Aiming for local solutions that can be cheaply computed, we seek to linearize the objective function. In a similar manner to (5.16), we write

$$\begin{aligned} \mathcal{E}^{+}_{\text{dft}}(\mathbf{z}) &= -\sum_{i=1}^{L} \frac{g_i \tau^2}{\pi e} \left\{\mathbb{E}\left[\det\left(\mathbf{I}_{g_i} + \gamma \mathbf{H}_i^{\text{H}} \mathbf{T} \mathbf{Z}_i \mathbf{T}^{\text{H}} \mathbf{H}_i\right)\right]\right\}^{-1/g_i} \\ &\le -\sum_{i=1}^{L} \underbrace{\frac{g_i \tau^2}{\pi e} \left(1 + \gamma \mathbf{a}_i^{\text{T}} \mathbf{z}_i\right)^{-1}}_{\triangleq f_i(\mathbf{z}_i)}, \end{aligned}$$

where by definition $\mathbf{a}_i \triangleq \text{diag}(\mathbf{T}^{\text{H}} \mathbf{\Sigma}_i \mathbf{T})$. In view of the convexity of $f_i(\mathbf{z}_i)$, a linear under-estimator can be obtained according to

$$f_i\left(\mathbf{z}_i\right) \ge f_i\left(\bar{\mathbf{z}}_i\right) + \nabla^{\text{T}} f_i\left(\bar{\mathbf{z}}_i\right)\left(\mathbf{z}_i - \bar{\mathbf{z}}_i\right),\ 1 \le i \le L,$$

where $\bar{\mathbf{z}}_i \in \mathbb{B}^{N_{\text{t}}}$ is chosen as a minimizer of $f_i(\mathbf{z}_i)$ subject to the feasibility constraint, i.e., it picks the g_i largest elements of $\mathbf{a}_i$, and the corresponding gradient $\nabla f_i(\bar{\mathbf{z}}_i)$ is evaluated as $\nabla f\left(\bar{\mathbf{z}}_i\right) = -\frac{g_i \tau^2 \gamma}{\pi e \left(1 + \gamma \mathbf{a}_i^{\text{T}} \bar{\mathbf{z}}_i\right)^2} \mathbf{a}_i$. In combination with $\mathcal{E}^{-}_{\text{dft}}(\mathbf{z})$, we arrive at a linear over-estimator for $\mathcal{E}^{\text{pcp}}_{\text{dft}}(\mathbf{z})$, which is solved for a local feasible solution, i.e.,

$$\max_{\{\mathbf{z} \in \mathbb{B}^{LN_{\text{t}}} \mid \mathbf{z} \in [\bar{\mathbf{p}}, \bar{\mathbf{q}}]\}} \quad \hat{\mathcal{E}}^{\text{pcp}}_{\text{dft}} \triangleq -\sum_{i=1}^{L} \left[f_i\left(\bar{\mathbf{z}}_i\right) + \nabla^{\text{T}} f_i\left(\bar{\mathbf{z}}_i\right)\left(\mathbf{z}_i - \bar{\mathbf{z}}_i\right) + \mathbf{b}_i^{\text{T}} \mathbf{z}_i\right]$$

$$\text{s.t.} \qquad h_{-}(\mathbf{z}) \leq 0 \leq h_{+}(\mathbf{z}), \tag{5.25}$$

with $\mathbf{b}_i \triangleq \operatorname{diag}(\mathbf{T}^{\mathrm{H}}\overline{\boldsymbol{\Sigma}}_i\mathbf{T})$. Although non-convex, such a mixed-integer linear program can be approximately solved with good efficiency by off-the-shelf software packages, e.g., cvx [19]. Let $\mathbf{z}_{\mathrm{LP}}$ denote a solution to (5.25). The resulting objective value can thus be used as an alternative to a local upper bound, i.e., $\mu([\bar{\mathbf{p}}, \bar{\mathbf{q}}]) = \min\left\{\mathcal{E}_{\mathrm{dft}}^{+}(\bar{\mathbf{q}}) - \mathcal{E}_{\mathrm{dft}}^{-}(\bar{\mathbf{p}}), \hat{\mathcal{E}}_{\mathrm{dft}}^{\mathrm{pcp}}(\mathbf{z}_{\mathrm{LP}})\right\}$. By leveraging the framework in Algorithm 5.3, the problem can be readily solved to global optimality.

5.5 Illustrative Results and Discussions

In this section, we numerically evaluate the error performance of the proposed nonlinear hybrid solutions in terms of bit error rate (BER), which is plotted versus SNR per user, defined as $\mathrm{SNR} = \frac{P_{\mathrm{t}}}{K\sigma_n^2}$. For the single-cluster scenario, we are interested to know to what extent introducing VP offsets performance loss attributed to limited RF chains and partial CSI in the RF domain. For the multi-cluster scenario, we assess the effectiveness of our joint approach to combining RF precoding with baseband MMSE-VP.

5.5.1 *Simulation Setup*

The simulations are set up as follows. The BS is equipped with $N_{\mathrm{t}} = 64$ transmit antennas driven by $n_{\mathrm{t}} = K$ RF chains. For the single-cluster scenario, $K = 8$ users are clustered at the same hotspot with a AoD $\bar{\theta} = 0°$ and angle spread $\Delta = 6°$. For the multi-cluster scenario, $K = 9$ users are separated into $L = 3$ clusters, each with three users, i.e., $g_i = 3, i = 1, 2, 3$. The one-ring channel correlation model in (6.1) is generated with mean AoDs $\bar{\theta}_1 = -10°, \bar{\theta}_2 = 0°, \bar{\theta}_3 = 10°$ and a common angle spread $\Delta = 6°$. In this setting, there is considerable overlap between channel power azimuth spectra, which results in strong inter-cluster interference. 16-QAM modulation is used for transmission, and the parameter τ is set as four times the minimum distance between constellation symbols for modulo decoding. In Algorithm 2, the RF precoder is initialized as the eigenvectors corresponding to the n_{t} largest eigenvalues of $\sum_{i=1}^{L} \boldsymbol{\Sigma}_i$, where we recall that $\boldsymbol{\Sigma}_i$ is the transmit correlation matrix for the ith user cluster, and the convergence threshold is set as $\epsilon = 10^{-6}$. The duration of the channel coherence block is assumed as $T = 100$ channel uses, and the numerical results are averaged over 5000 channel realizations.

5.5.2 Single-Cluster Scenario

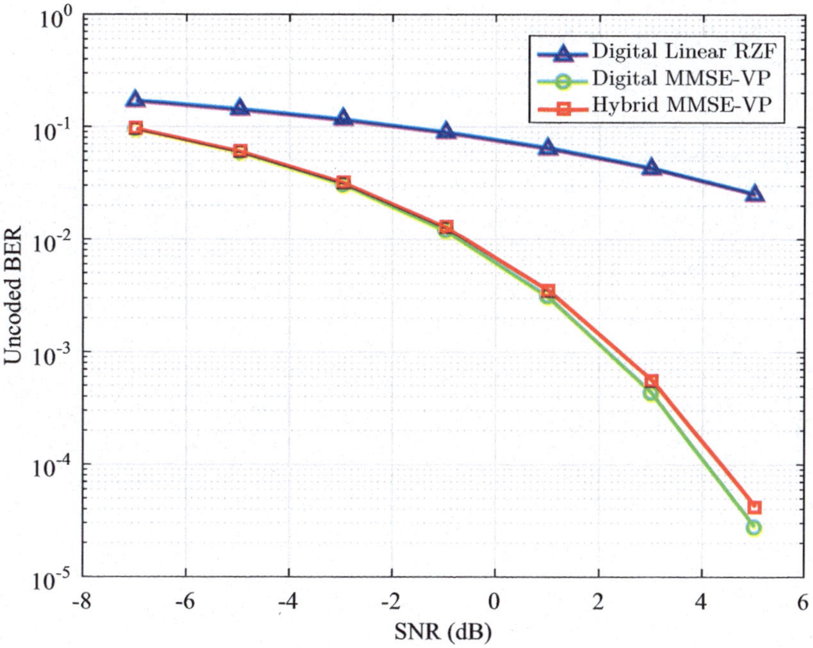

Fig. 5.2 BERs achieved by linear and nonlinear precoding schemes for a 64×8 massive MIMO system with 8 RF chains at the transmitter. The users are clustered at a hotspot. The mean AoD is 0°, and the angle spread is 6°

In Fig. 5.2, we evaluate the error performance of the proposed nonlinear hybrid precoder, i.e., RF eigen-beamforming with baseband MMSE-VP (denoted as Hybrid MMSE-VP) when only a single cluster of users are scheduled for transmission. The fully digital linear RZF precoder (denoted as Digital linear RZF) [2] and the traditional MMSE-VP precoder (denoted as Digital MMSE-VP) [10], which are perfect CSI-based, are also presented for comparison. Remarkably, with only a limited number of RF chains ($n_{\mathrm{t}} = K = 8$), the two-timescale hybrid solution enjoys substantial performance gain over the perfect RZF solution while suffering a fairly small power loss compared with its perfect counterpart. We note that in a propagation environment with limited local scattering, as represented by a small degree of angle spread ($\Delta = 6°$), channel power tends to concentrate in the eigen-domain. This translates to a limited degree of spatial multiplexing supported by the channel. When the system is fully loaded, not surprisingly, the linear scheme suffers severe power loss. Another consequence is that the power gain of the eigen-domain is largely preserved even when the high dimension of the MIMO channel is significantly reduced. This accounts for the small performance gap between Hybrid MMSE-VP and Digital MMSE-VP.

5.5.3 Multi-Cluster Scenario

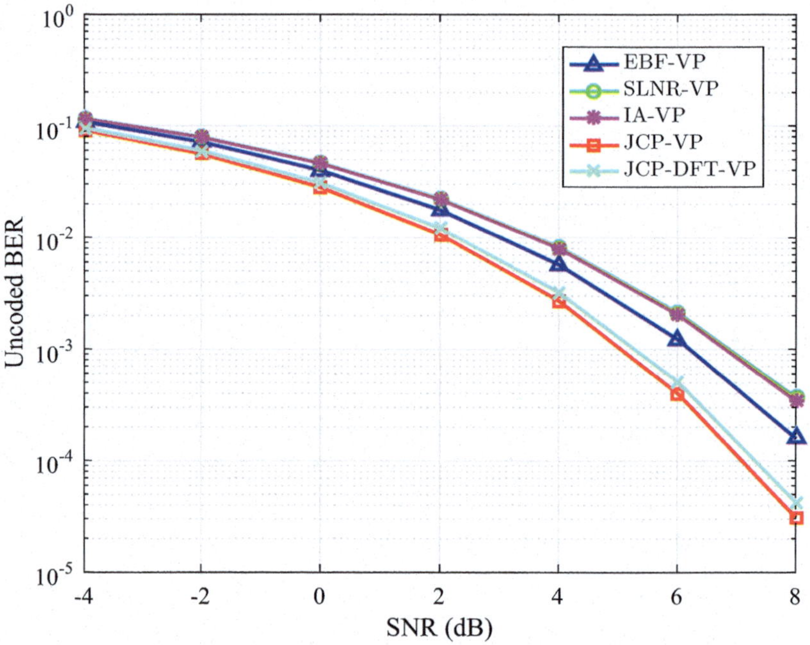

Fig. 5.3 BERs achieved by various hybrid precoding schemes with joint MMSE-VP for a 64×9 massive MIMO system with 9 RF chains at the transmitter. The users are clustered at three hotspots, each with three users. The mean AoDs are $-10°$, $0°$, $10°$, and the angle spread is $6°$

In Figs. 5.3 and 5.5, we plot the BERs of the proposed joint design of RF precoding with joint baseband MMSE-VP (denoted as JCP-VP) and cluster-wise baseband MMSE-VP (denoted as PCP-VP), respectively. A comparison is drawn with three other state-of-the-art two-timescale hybrid precoding schemes in the literature, i.e., (1) the heuristic statistical eigen-beamforming precoder (denoted as EBF-VP) [9, 20]; (2) the signal-to-leakage-plus-noise ratio (SLNR)-based precoder (denoted as SLNR-VP) [21]; (3) the interference alignment-based precoder (denoted as IA-VP) [22]. For fairness, MMSE-VP precoding is employed in place of the original linear counterparts at baseband.

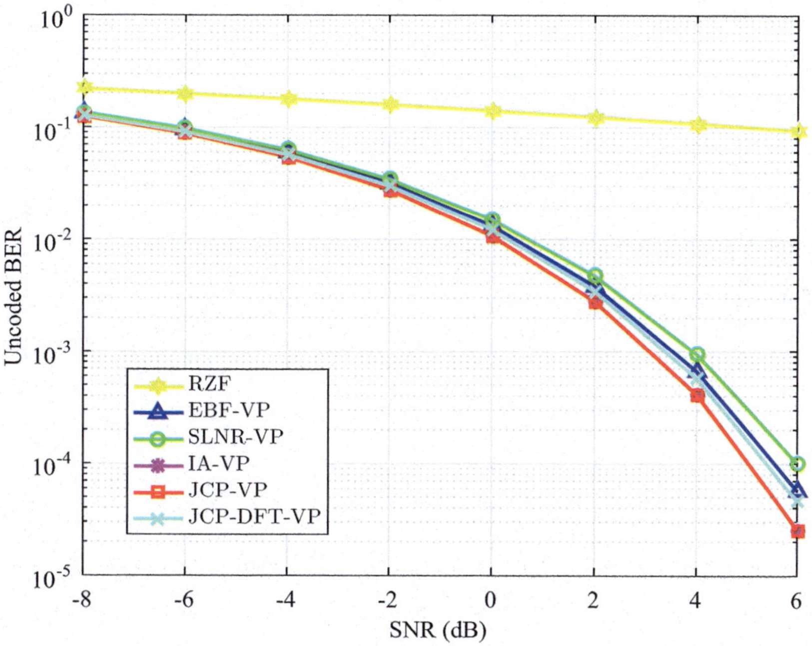

Fig. 5.4 BERs achieved by various hybrid precoding schemes with joint MMSE-VP for a 64×16 massive MIMO system with 16 RF chains at the transmitter. The users are clustered at two hotspots, each with eight users. The mean AoDs are $-10°$, $10°$, and the angle spread is $6°$

For the case of joint baseband processing in Figs. 5.3 and 5.4, it is seen that the proposed hybrid precoder noticeably outperforms the other three baselines. Intuitively, the overall system performance benefits from a reasonable trade-off between self-transmission and interference control for each cluster. With perfect effective CSI at baseband and partial CSI at RF, such conflicting objectives should also be balanced in the short term and the long term. This could only be achieved by jointly considering the RF stage and the nonlinear baseband stage. We note that since joint baseband processing enables the baseband to fine-tune the overall RF beam domain across clusters and therefore delivers a satisfactory error performance in the medium SNR regime, separate design of RF and baseband precoding does not exhibit interference-limited behaviors. However, this is no longer the case when baseband precoding has only access to the cluster-wise RF beam domain, as shown in Fig. 5.5. Because EBF-VP over-penalizes self-transmission by forcing inter-cluster interference to zero, severe power loss occurs. On the other hand, although SLNR-VP and IA-VP aim for a long-term trade-off between self-transmission optimization and interference leakage power management, they fail to coordinate with the nonlinear baseband, and hence become interference-limited in the medium-to-high SNR regime.

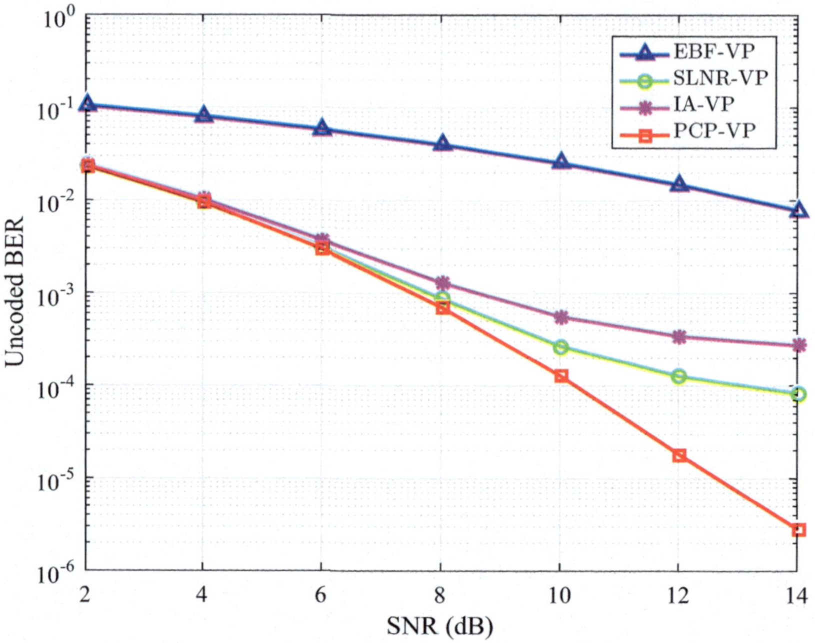

Fig. 5.5 BERs achieved by various hybrid precoding schemes with cluster-wise MMSE-VP for a 64×9 massive MIMO system with 9 RF chains at the transmitter. The users are clustered at three hotspots, each with three users. The mean AoDs are $-10°$, $0°$, $10°$, and the angle spread is $6°$

Because DFT codebooks represent uniform quantization of the angular domain, the spatial resolution in terms of beam directions and beam-widths is limited by the array dimension. Nonetheless, when combined with joint MMSE-VP (denoted as JCP-DFT-VP), such a constant-modulus alternative delivers a competitive error performance against the nonuniform-modulus counterpart, as illustrated in Fig. 5.3. This is attributed to the flexibility of *joint* baseband precoding in adjusting the RF beam domain. Unfortunately, when baseband precoding is restricted to each cluster, the restrictive DFT beams render it ineffective in achieving cluster separation without excessively penalizing self-transmission. The performance degradation is especially appreciable for limited scattering (e.g., $\Delta = 6°$) where a limited set of beam choices are available, as dictated by (5.14). Naturally, when a larger antenna array is deployed in a rich scattering environment, such an issue is expected to be mitigated. As illustrated in Fig. 5.6, DFT beamforming with $N_\mathrm{t} = 128$ elements (denoted as PCP-DFT-VP) performs comparably with the other nonuniform-modulus precoding solutions in the presence of moderate inter-cluster interference and angle spread $\Delta = 18°$.

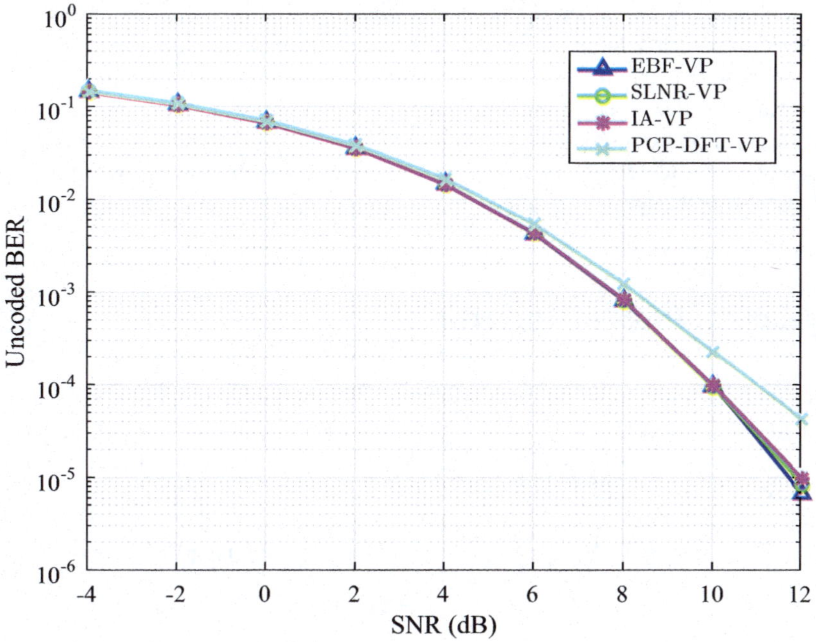

Fig. 5.6 BERs achieved by various hybrid precoding schemes with cluster-wise MMSE-VP for a 128×6 massive MIMO system with 8 RF chains at the transmitter. The users are clustered at two hotspots, each with three users. The mean AoDs are $\bar{\theta}_1 = 0°$ and $\bar{\theta}_2 = 40°$, and the angle spread is 18°

5.6 Summary

In this chapter, we investigated MSE based two-timescale nonlinear hybrid precoder design for massive MU-MIMO systems. A two-step approach to joint RF-baseband optimization was proposed, where the statistical CSI-based RF precoding with nonuniform- and constant-modulus elements was addressed in respect of the instantaneous effective CSI-based joint MMSE-VP and cluster-wise MMSE-VP at baseband. For the single-cluster scenario, statistical eigen-beamforming in the RF domain was proved to be optimal. For the multi-cluster scenario, in that there was no closed-form characterization of the objective functions as a result of the nonlinear VP search, we proposed mathematically tractable lower bounds as approximations of the objective functions. The nonuniform-envelope RF designs were recognized as numerical optimization problems on the smooth Riemannian manifolds, and solved by trust-region Newton methods with global convergence at a locally superlinear rate. The constant-modulus RF design was simplified as the DFT codebook-based beam selection, which was efficiently solved to global optimality by exploiting the framework of discrete monotonic optimization. Simulation results showed that for the single-cluster scenarios with limited scattering, introducing MMSE-

VP compensated for the performance penalty incurred by limited RF chains and partial CSI compared with its fully digital linear counterpart. For the multi-cluster scenarios where strong inter-cluster interference was present, the proposed approach of joint design delivered noticeably superior error performance to other state-of-the-art hybrid precoding solutions. In addition, DFT beam selection was shown to be an effective alternative that achieved a good trade-off between performance and hardware complexity.

Appendix 1: Proof of Theorem 5.1

Recall that $\boldsymbol{\Sigma}_i = \mathbf{U}_i \boldsymbol{\Lambda}_i \mathbf{U}_i^{\mathrm{H}}$ with the diagonal elements $\lambda_{i,1}, \ldots, \lambda_{i,r_i}$ of $\boldsymbol{\Lambda}_i$ in non-increasing order. Let $\mathbf{U}_{i,1} \in \mathbb{C}^{N_{\mathrm{t}} \times n_{\mathrm{t}}}$ be a matrix whose columns are the eigenvectors associated with the n_{t} largest eigenvalues $\lambda_{i,1}, \ldots, \lambda_{i,n_{\mathrm{t}}}$. By spectral decomposition, we rewrite

$$
\begin{aligned}
(\boldsymbol{\Sigma}_i^{1/2})^{\mathrm{H}} \mathbf{F}_{\mathrm{RF}} \mathbf{F}_{\mathrm{RF}}^{\mathrm{H}} \boldsymbol{\Sigma}_i^{1/2} &= \boldsymbol{\Lambda}_i^{1/2} \mathbf{U}_i^{\mathrm{H}} \mathbf{F}_{\mathrm{RF}} \mathbf{F}_{\mathrm{RF}}^{\mathrm{H}} \mathbf{U}_i \boldsymbol{\Lambda}_i^{1/2} \\
&= \mathbf{V}\mathbf{D}\mathbf{V}^{\mathrm{H}},
\end{aligned}
$$

where $\mathbf{V}$ is unitary and the diagonal elements of $\mathbf{D}$ are in non-increasing order. Using the fact that $\mathbf{AB}$ and $\mathbf{BA}$ have the same nonzero eigenvalues, we conclude that the nonzero eigenvalues of $\mathbf{F}_{\mathrm{RF}}^{\mathrm{H}} \mathbf{U}_i \boldsymbol{\Lambda}_i \mathbf{U}_i^{\mathrm{H}} \mathbf{F}_{\mathrm{RF}}$ are the same as those of $\mathbf{D}$. Furthermore, since $\mathbf{F}_{\mathrm{RF}}^{\mathrm{H}} \mathbf{U}_i \in \mathbb{C}^{n_{\mathrm{t}} \times r_i}$ is semi-unitary, the Poincaré separation theorem [23] indicates that the maximum elements that $\mathbf{D}$ could have on the diagonal are the maximum n_{t} diagonal elements of $\boldsymbol{\Lambda}_i$, i.e., $\boldsymbol{\Lambda}_{i,1} \triangleq \mathrm{diag}\{\lambda_{i,1}, \ldots, \lambda_{i,n_{\mathrm{t}}}\} \succeq \mathbf{D}$, and this can be achieved by setting $\mathbf{F}_{\mathrm{RF}} = \mathbf{U}_{i,1}$. Let $\mathbf{s}_i^{[t]}(\mathbf{F}_{\mathrm{RF}}) \triangleq \mathbf{d}[t] + \tau \mathbf{q}^{[t]}(\mathbf{F}_{\mathrm{RF}})$ be the signal perturbed by the MMSE perturbation vector $\mathbf{q}^{[t]}(\mathbf{F}_{\mathrm{RF}})$ corresponding to the RF precoder $\mathbf{F}_{\mathrm{RF}}$. Using the previous observations, we may establish that

$$
\begin{aligned}
&\mathbb{E}_{\mathbf{H}_{i,w}, \mathbf{d}_i[t]} \left[\mathbf{s}_i^{[t]}(\mathbf{F}_{\mathrm{RF}})^{\mathrm{H}} (\mathbf{H}_i^{\mathrm{H}} \mathbf{F}_{\mathrm{RF}} \mathbf{F}_{\mathrm{RF}}^{\mathrm{H}} \mathbf{H}_i + \gamma \mathbf{I}_{g_i})^{-1} \mathbf{s}_i^{[t]}(\mathbf{F}_{\mathrm{RF}}) \right] \\
\overset{(a)}{=} \; &\mathbb{E}_{\mathbf{H}_{i,w}, \mathbf{d}_i[t]} \left[\mathbf{s}_i^{[t]}(\mathbf{F}_{\mathrm{RF}})^{\mathrm{H}} \left(\mathbf{H}_{i,w}^{\mathrm{H}} \mathbf{D} \mathbf{H}_{i,w} + \gamma \mathbf{I}_{g_i} \right)^{-1} \mathbf{s}_i^{[t]}(\mathbf{F}_{\mathrm{RF}}) \right] \\
\overset{(b)}{\geq} \; &\mathbb{E}_{\mathbf{H}_{i,w}, \mathbf{d}_i[t]} \left[\mathbf{s}_i^{[t]}(\mathbf{F}_{\mathrm{RF}})^{\mathrm{H}} \left(\mathbf{H}_{i,w}^{\mathrm{H}} \boldsymbol{\Lambda}_{i,1} \mathbf{H}_{i,w} + \gamma \mathbf{I}_{g_i} \right)^{-1} \mathbf{s}_i^{[t]}(\mathbf{F}_{\mathrm{RF}}) \right] \\
\overset{(c)}{\geq} \; &\mathbb{E}_{\mathbf{H}_{i,w}, \mathbf{d}_i[t]} \left[\mathbf{s}_i^{[t]}(\mathbf{U}_{i,1})^{\mathrm{H}} \left(\mathbf{H}_{i,w}^{\mathrm{H}} \boldsymbol{\Lambda}_{i,1} \mathbf{H}_{i,w} + \gamma \mathbf{I}_{g_i} \right)^{-1} \mathbf{s}_i^{[t]}(\mathbf{U}_{i,1}) \right],
\end{aligned}
$$

where (a) is due to the well-known bi-unitary invariance property of Gaussian distribution, (b) is a consequence of $\boldsymbol{\Lambda}_{i,1} \succeq \mathbf{D}$, and the optimality of $\mathbf{q}^{[t]}(\mathbf{U}_{i,1})$ implies (c). In summary, $\mathbf{U}_{i,1}$ indeed achieves MMSE for single-cluster transmission.

Appendix 2: Proof of Lemma 5.1

Note that the MSE $\mathcal{E}_{\text{jcp}}^{\star}(\mathbf{F}_{\text{RF}})$ associated with the optimal perturbation vector is nothing but the LS errors resulting from (5.4). Under the assumption of uniform distribution, a direct consequence from the lattice quantization theory is that such errors, averaged over all possible inputs, can be lower-bounded by [13, 24, 25]

$$\mathbb{E}_{\{\mathbf{H}_{i,w},\mathbf{d}_i\}}\left[\mathcal{E}_{\text{jcp}}^{\star}(\mathbf{F}_{\text{RF}})\right] \geq \frac{K\tau^2}{\pi e}\mathbb{E}_{\{\mathbf{H}_{i,w},\mathbf{d}_i\}}\left\{\left[\det(\mathbf{I}_K + \gamma\mathbf{F}_{\text{RF}}^{\text{H}}\mathbf{H}\mathbf{H}^{\text{H}}\mathbf{F}_{\text{RF}})\right]^{-\frac{1}{K}}\right\}.$$

Using the identity $\det(\mathbf{I}+\mathbf{AB}) = \det(\mathbf{I}+\mathbf{BA})$ and applying Jensen's inequality, we arrive at the lower bound $\mathcal{E}_{\text{jcp,lb}}(\mathbf{F}_{\text{RF}})$.

Appendix 3: Proof of Theorem 5.2

Recall that $\mathbf{H} = \boldsymbol{\Sigma}^{1/2}\mathbf{H}_w$ with $\mathbf{H}_w = \text{diag}\{\mathbf{H}_{1,w},\ldots,\mathbf{H}_{L,w}\}$. Applying the principal minor determinant expansion, we have

$$\begin{aligned}
\mathcal{J}(\mathbf{F}_{\text{RF}}) &= \mathbb{E}\left[\det\left(\mathbf{I}_K + \gamma\mathbf{H}_w^{\text{H}}(\boldsymbol{\Sigma}^{1/2})^{\text{H}}\mathbf{F}_{\text{RF}}\mathbf{F}_{\text{RF}}^{\text{H}}\boldsymbol{\Sigma}^{1/2}\mathbf{H}_w\right)\right] \\
&= \sum_{k=0}^{K}\gamma^k \sum_{1\leq i_1<\cdots<i_k\leq K}\mathbb{E}\left\{\det\left[\left(\mathbf{H}_w^{\text{H}}(\boldsymbol{\Sigma}^{1/2})^{\text{H}}\mathbf{F}_{\text{RF}}\mathbf{F}_{\text{RF}}^{\text{H}}\boldsymbol{\Sigma}^{1/2}\mathbf{H}_w\right)_{i_1,\ldots,i_k}^{i_1,\ldots,i_k}\right]\right\} \\
&= \sum_{k=0}^{K}\gamma^k \sum_{\substack{1\leq j_1<\cdots<j_k\leq LN_{\text{t}} \\ 1\leq u_1<\cdots<u_k\leq LN_{\text{t}}}} \det\left[\left((\boldsymbol{\Sigma}^{1/2})^{\text{H}}\mathbf{F}_{\text{RF}}\mathbf{F}_{\text{RF}}^{\text{H}}\boldsymbol{\Sigma}^{1/2}\right)_{u_1,\ldots,u_k}^{j_1,\ldots,j_k}\right] \\
&\quad\times \sum_{1\leq i_1<\cdots<i_k\leq K}\mathbb{E}\left[\det\left(\left(\mathbf{H}_w^{\text{H}}\right)_{j_1,\ldots,j_k}^{i_1,\ldots,i_k}\right)\det\left((\mathbf{H}_w)_{i_1,\ldots,i_k}^{u_1,\ldots,u_k}\right)\right]
\end{aligned}$$

where the last step results from the Cauchy-Binet formula [26]. By the definition of determinant, the expectation term can be explicitly evaluated as

$$\begin{aligned}
&\mathbb{E}\left[\det\left(\left(\mathbf{H}_w^{\text{H}}\right)_{j_1,\ldots,j_k}^{i_1,\ldots,i_k}\right)\det\left((\mathbf{H}_w)_{i_1,\ldots,i_k}^{u_1,\ldots,u_k}\right)\right] \\
&\qquad = \begin{cases}\prod_{s=1}^{L}\frac{g_s!}{(g_s-|\mathcal{I}_{i,k}|)!}, & \text{if } j_m = u_m, \pi_m = \rho_m, 1\leq m\leq k \\ 0, & \text{otherwise}\end{cases},
\end{aligned}$$

where we recall that $\mathcal{I}_{i,k} = \{j_{i_1},\ldots,j_{i_k}\}$ denotes the number of rows chosen from $\mathbf{H}_{i,w}$.

Appendix 4: Derivation of Riemannian Gradient and Riemannian Hessian

Joint Hybrid MMSE-VP Precoding

The Riemannian gradient is readily obtainable once its Euclidean counterpart is known. From the Fréchet differential of $\mathcal{J}(\mathbf{F}_{\mathrm{RF}})$ along the direction $\mathbf{Z}$, i.e.,

$$\mathcal{D}\mathcal{J}\left(\mathbf{F}_{\mathrm{RF}}\right)[\mathbf{Z}] = \frac{d}{dt}\mathcal{J}\left(\mathbf{F}_{\mathrm{RF}} + t\mathbf{Z}\right)\Big|_{t=0},$$

the Euclidean gradient is identified as

$$\nabla\mathcal{J}\left(\mathbf{F}_{\mathrm{RF}}\right) = 2\sum_{k=1}^{K}\gamma^k \sum_{\mathcal{I}_k\subset[\sum_i r_i]} \alpha_k\left(\mathcal{I}_k\right)\det\left(\boldsymbol{\Phi}_{\mathcal{I}_k}\right)\boldsymbol{\Xi}_{\mathcal{I}_k}\mathbf{F}_{\mathrm{RF}},$$

where the identities

$$\frac{d}{dt}\mathbf{A}(t)^{-1} = -\mathbf{A}(t)^{-1}\cdot\left[\frac{d}{dt}\mathbf{A}(t)\right]\cdot\mathbf{A}(t)^{-1},$$

$$\frac{d}{dt}\det\left[\mathbf{A}(t)\right] = \det\left(\mathbf{A}(t)\right)\mathrm{tr}\left(\left(\mathbf{A}(t)\right)^{-1}\frac{d}{dt}\mathbf{A}(t)\right)$$

are used, and $\boldsymbol{\Xi}_{\mathcal{I}_k} \triangleq \boldsymbol{\Sigma}_{\mathcal{I}_k}^{1/2}\boldsymbol{\Phi}_{\mathcal{I}_k}^{-1}(\boldsymbol{\Sigma}_{\mathcal{I}_k}^{1/2})^{\mathrm{H}}$. Accordingly, the projection of $\nabla\mathcal{J}\left(\mathbf{F}_{\mathrm{RF}}\right)$ onto the tangent space $\mathcal{T}_{\mathbf{F}_{\mathrm{RF}}}\mathcal{M}_{\mathrm{gr}}$, i.e., $\mathcal{P}_{\mathbf{F}_{\mathrm{RF}}}^{\mathrm{gr}}\nabla\mathcal{J}\left(\mathbf{F}_{\mathrm{RF}}\right)$, yields the Riemannian gradient grad $\mathcal{J}\left(\mathbf{F}_{\mathrm{RF}}\right)$ in (5.10) [14].

To derive the Riemannian Hessian, we start with the definition of geodesic. By compact SVD, a tangent vector $\boldsymbol{\eta} \in \mathcal{T}_{\mathbf{F}_{\mathrm{RF}}}\mathcal{M}_{\mathrm{gr}}$ is written as $\boldsymbol{\eta} = \mathbf{U}_{\eta}\boldsymbol{\Sigma}_{\eta}\mathbf{V}_{\eta}^{\mathrm{H}}$. Accordingly, a geodesic $\upsilon_{\mathbf{F}_{\mathrm{RF}}}(t)$ originating from $\mathbf{F}_{\mathrm{RF}}$ is expressed as $\upsilon_{\mathbf{F}}(t) = \mathbf{F}_{\mathrm{RF}}\mathbf{V}_{\eta}\cos(\boldsymbol{\Sigma}_{\eta}t)\mathbf{V}_{\eta}^{\mathrm{H}} + \mathbf{U}_{\eta}\sin(\boldsymbol{\Sigma}_{\eta}t)\mathbf{V}_{\eta}^{\mathrm{H}}$ [27] and satisfies

$$\upsilon_{\mathbf{F}}(0) = \mathbf{F}_{\mathrm{RF}},\quad \frac{d\upsilon_{\mathbf{F}}(t)}{dt} = \boldsymbol{\eta},\quad \frac{d^2\upsilon_{\mathbf{F}}(t)}{dt^2} + \upsilon_{\mathbf{F}}(t)\left[\frac{d\upsilon_{\mathbf{F}}(t)^{\mathrm{H}}}{dt}\frac{d\upsilon_{\mathbf{F}}(t)}{dt}\right] = 0.$$

The Riemannian Hessian is related to $\upsilon_{\mathbf{F}}(t)$ by

$$\Re\mathrm{tr}\left(\boldsymbol{\eta}^{\mathrm{H}}\mathrm{Hess}\,\mathcal{J}\left(\mathbf{F}_{\mathrm{RF}}\right)[\boldsymbol{\eta}]\right) = \frac{d^2}{dt^2}\mathcal{J}\left(\upsilon_{\mathbf{F}}(t)\right).$$

By evaluating explicitly the second-order derivative on the right-hand side, it can be shown that $\Re\{\mathrm{tr}(\boldsymbol{\eta}^{\mathrm{H}}\mathrm{Hess}\,\mathcal{J}\left(\mathbf{F}_{\mathrm{RF}}\right)[\boldsymbol{\eta}])\}$ is quadratic in $\boldsymbol{\eta}$. Based on the polarization identity [14], it can be further established that

$$\begin{aligned}&\Re\mathrm{tr}\left(\boldsymbol{\xi}^{\mathrm{H}}\mathrm{Hess}\,\mathcal{J}\left(\mathbf{F}_{\mathrm{RF}}\right)[\boldsymbol{\eta}]\right)\\&=2\sum_{k=1}^{K}\gamma^{k}\sum_{\mathcal{I}_k\subset[\sum_i r_i]}\alpha_k\left(\mathcal{I}_k\right)\det\left(\boldsymbol{\Phi}_{\mathcal{I}_k}\right)\mathrm{tr}\left[\boldsymbol{\Xi}_{\mathcal{I}_k}\left(\boldsymbol{\eta}\mathbf{F}_{\mathrm{RF}}^{\mathrm{H}}+\mathbf{F}_{\mathrm{RF}}\boldsymbol{\eta}^{\mathrm{H}}\right)\right]\\&\times\Re\mathrm{tr}\left\{\boldsymbol{\xi}^{\mathrm{H}}\left[\boldsymbol{\Xi}_{\mathcal{I}_k}\mathbf{F}_{\mathrm{RF}}-\boldsymbol{\Xi}_{\mathcal{I}_k}\left(\boldsymbol{\eta}\mathbf{F}_{\mathrm{RF}}^{\mathrm{H}}+\mathbf{F}_{\mathrm{RF}}\boldsymbol{\eta}^{\mathrm{H}}\right)\boldsymbol{\Xi}_{\mathcal{I}_k}\mathbf{F}_{\mathrm{RF}}+\boldsymbol{\Xi}_{\mathcal{I}_k}\boldsymbol{\eta}-\boldsymbol{\eta}\mathbf{F}_{\mathrm{RF}}^{\mathrm{H}}\boldsymbol{\Xi}_{\mathcal{I}_k}\mathbf{F}_{\mathrm{RF}}\right]\right\},\end{aligned}$$

from which the Riemannian Hessian is identified as in (5.11).

Cluster-Wise Hybrid MMSE-VP Precoding

In a similar manner to the case of joint MMSE-VP, we start with evaluating the first-order differential along the direction $\mathbf{Z}$, i.e., $\mathcal{D}\mathcal{E}_{\mathrm{pcp,lb}}\left(\mathbf{F}_{\mathrm{RF}}\right)[\mathbf{Z}]=\frac{d}{dt}\mathcal{E}_{\mathrm{pcp,lb}}\left(\mathbf{F}_{\mathrm{RF}}+t\mathbf{Z}\right)\Big|_{t=0}$, which gives the Euclidean gradient as

$$\begin{aligned}&\nabla\mathcal{E}_{\mathrm{pcp,lb}}\left(\mathbf{F}_{\mathrm{RF}}\right)\\&=2\sum_{i=1}^{L}\left[-\frac{\tau^2}{\pi e}\tilde{\mathcal{E}}_i^{-(1+1/g_i)}\sum_{k=1}^{g_i}\gamma^k\sum_{\mathcal{I}_k\subset[r_i]}\alpha_{i,k}\det\left(\boldsymbol{\Phi}_{i,\mathcal{I}_k}\right)\boldsymbol{\Xi}_{i,\mathcal{I}_k}\mathbf{F}_{\mathrm{RF}}\mathbf{D}_i+\overline{\boldsymbol{\Sigma}}_i\mathbf{F}_{\mathrm{RF}}\mathbf{D}_i\right],\end{aligned}$$

where $\tilde{\mathcal{E}}_i\triangleq\sum_{k=0}^{g_i}\gamma^k\sum_{\mathcal{I}_k\subset[r_i]}\alpha_{i,k}\det(\boldsymbol{\Phi}_{i,\mathcal{I}_k})$, $\boldsymbol{\Xi}_{i,\mathcal{I}_k}\triangleq\boldsymbol{\Sigma}_{i,\mathcal{I}_k}^{1/2}\boldsymbol{\Phi}_{i,\mathcal{I}_k}^{-1}(\boldsymbol{\Sigma}_{i,\mathcal{I}_k}^{1/2})^{\mathrm{H}}$, and $\mathbf{D}_i\triangleq\mathbf{S}_i^{\mathrm{T}}\mathbf{S}_i$ with the ith selection matrix $\mathbf{S}_i\in\mathbb{B}^{n_i\times n_i}$ defined through $\mathbf{F}_{\mathrm{RF},i}=\mathbf{F}_{\mathrm{RF}}\mathbf{S}_i$ for $1\le i\le L$. Upon applying the orthogonal projector onto $\mathcal{T}_{\mathbf{F}_{\mathrm{RF}}}\mathcal{M}_{\mathrm{st}}$, i.e., $\mathcal{P}_{\mathbf{F}_{\mathrm{RF}}}^{\mathrm{st}}\mathbf{U}\triangleq\mathbf{U}-\mathbf{F}_{\mathrm{RF}}\frac{1}{2}\left(\mathbf{F}_{\mathrm{RF}}^{\mathrm{H}}\mathbf{U}+\mathbf{U}^{\mathrm{H}}\mathbf{F}_{\mathrm{RF}}\right)$, we have the Riemannian gradient as [14]

$$\begin{aligned}&\mathrm{grad}\,\mathcal{E}_{\mathrm{pcp,lb}}\left(\mathbf{F}_{\mathrm{RF}}\right)\\&=\mathcal{P}_{\mathbf{F}_{\mathrm{RF}}}^{\mathrm{st}}\nabla\mathcal{E}_{\mathrm{pcp,lb}}\left(\mathbf{F}_{\mathrm{RF}}\right)\\&=\nabla\mathcal{E}_{\mathrm{pcp,lb}}\left(\mathbf{F}_{\mathrm{RF}}\right)-\mathbf{F}_{\mathrm{RF}}\frac{1}{2}\left(\mathbf{F}_{\mathrm{RF}}^{\mathrm{H}}\nabla\mathcal{E}_{\mathrm{pcp,lb}}\left(\mathbf{F}_{\mathrm{RF}}\right)+\nabla^{\mathrm{H}}\mathcal{E}_{\mathrm{pcp,lb}}\left(\mathbf{F}_{\mathrm{RF}}\right)\mathbf{F}_{\mathrm{RF}}\right).\end{aligned}$$

Moreover, since the Euclidean Hessian is related to the Riemannian Hessian by the Weingarten map [18], it suffices to derive the former case, which is nothing but the directional derivative of the Euclidean gradient, i.e.,

$$\nabla^2\mathcal{E}_{\mathrm{pcp,lb}}\left(\mathbf{F}_{\mathrm{RF}}\right)[\boldsymbol{\eta}]=\frac{d}{dt}\nabla\mathcal{E}_{\mathrm{pcp,lb}}\left(\mathbf{F}_{\mathrm{RF}}+t\boldsymbol{\eta}\right)\Big|_{t=0}.$$

By direct evaluation, we arrive at

$$
\begin{aligned}
&\nabla^2 \mathcal{E}_{\mathrm{pcp,lb}}\left(\mathbf{F}_{\mathrm{RF}}\right)[\boldsymbol{\eta}] \\
&= 2\sum_{i=1}^{L} \overline{\boldsymbol{\Sigma}}_i \boldsymbol{\eta} \mathbf{D}_i + 2\sum_{i=1}^{L} \frac{\tau^2}{\pi e} \tilde{\mathcal{E}}_i^{-(1+1/g_i)} \sum_{k=1}^{g_i} \gamma^k \sum_{\mathcal{I}_k \subset [r_i]} \alpha_{i,k} \det\left(\boldsymbol{\Phi}_{i,\mathcal{I}_k}\right) \\
&\times \Bigg[2\left(1+g_i^{-1}\right) \tilde{\mathcal{E}}_i^{-1} \Re\left\{\mathrm{tr}\left(\boldsymbol{\eta}_i^{\mathrm{H}} \boldsymbol{\Xi}_{i,\mathcal{I}_k} \mathbf{F}_{\mathrm{RF},i}\right)\right\} \sum_{l=1}^{g_i} \gamma^l \sum_{\mathcal{I}_l \subset [r_i]} \alpha_{i,l} \det\left(\boldsymbol{\Phi}_{i,\mathcal{I}_l}\right) \boldsymbol{\Xi}_{i,\mathcal{I}_l} \mathbf{F}_{\mathrm{RF}} \mathbf{D}_i \\
&\qquad - 2\Re\left\{\mathrm{tr}\left[\boldsymbol{\eta}_i^{\mathrm{H}} \boldsymbol{\Xi}_{i,\mathcal{I}_k} \mathbf{F}_{\mathrm{RF},i}\right]\right\} \boldsymbol{\Xi}_{i,\mathcal{I}_k} \mathbf{F}_{\mathrm{RF}} \mathbf{D}_i - \boldsymbol{\Xi}_{i,\mathcal{I}_k} \boldsymbol{\eta} \mathbf{D}_i \\
&\qquad + \boldsymbol{\Xi}_{i,\mathcal{I}_k} \left(\boldsymbol{\eta}_i \mathbf{F}_{\mathrm{RF},i}^{\mathrm{H}} + \mathbf{F}_{\mathrm{RF},i} \boldsymbol{\eta}_i^{\mathrm{H}}\right) \boldsymbol{\Xi}_{i,\mathcal{I}_k} \mathbf{F}_{\mathrm{RF}} \mathbf{D}_i \Bigg].
\end{aligned}
$$

Therefore, the Riemannian Hessian is obtained as

$$
\begin{aligned}
&\mathrm{Hess}\, \mathcal{E}_{\mathrm{pcp,lb}}\left(\mathbf{F}_{\mathrm{RF}}\right)[\boldsymbol{\eta}] \\
&= \mathcal{P}_{\mathbf{F}_{\mathrm{RF}}}^{\mathrm{st}} \nabla^2 \mathcal{E}_{\mathrm{pcp,lb}}\left(\mathbf{F}_{\mathrm{RF}}\right)[\boldsymbol{\eta}] + \mathfrak{A}_{\mathbf{F}_{\mathrm{RF}}}^{\mathrm{st}}\left(\boldsymbol{\eta}, \nabla \mathcal{E}_{\mathrm{pcp,lb}}\left(\mathbf{F}_{\mathrm{RF}}\right) - \mathcal{P}_{\mathbf{F}_{\mathrm{RF}}}^{\mathrm{st}} \nabla \mathcal{E}_{\mathrm{pcp,lb}}\left(\mathbf{F}_{\mathrm{RF}}\right)\right)
\end{aligned}
$$

with the Weingarten map defined as [18]

$$
\mathfrak{A}_{\mathbf{F}_{\mathrm{RF}}}^{\mathrm{st}}\left(\mathbf{Z}, \mathbf{V}\right) \triangleq -\mathbf{Z}\mathbf{F}_{\mathrm{RF}}^{\mathrm{H}}\mathbf{V} - \frac{1}{2}\mathbf{F}_{\mathrm{RF}}\left(\mathbf{Z}^{\mathrm{H}}\mathbf{V} + \mathbf{V}^{\mathrm{H}}\mathbf{Z}\right).
$$

Appendix 5: Proof of Proposition 5.1

Smoothness follows directly from the fact that determinant is a multi-linear function of $z_1, \ldots, z_{N_t}$. To prove monotonicity, let $\bar{\mathbf{z}} \triangleq \mathbf{z} + \mathbf{e}_j$ with $[\mathbf{z}]_j = 0$ for some j, or, $\mathrm{diag}\{\bar{\mathbf{z}}\} = \mathbf{Z} + \mathbf{e}_j\mathbf{e}_j^{\mathrm{T}}$. Using the identity $\det(\mathbf{A} + \mathbf{x}\mathbf{x}^{\mathrm{H}}) = \det(\mathbf{A}) + \mathbf{x}^{\mathrm{H}}\mathrm{adj}(\mathbf{A})\mathbf{x}$, it is easy to show that

$$
\begin{aligned}
&\mathcal{J}_{\mathrm{dft}}\left(\bar{\mathbf{z}}\right) \\
&= \mathcal{J}_{\mathrm{dft}}\left(\mathbf{z}\right) + \sum_{k=1}^{K} \gamma^k \sum_{\mathcal{I}_k = \bigcup_{i=1}^{L} \mathcal{I}_{i,k} \subset [\sum_i r_i]} \alpha_k\left(\mathcal{I}_k\right) \mathbf{t}_j^{\mathrm{H}} \boldsymbol{\Sigma}_{\mathcal{I}_k}^{1/2} \mathrm{adj}\left(\boldsymbol{\Phi}_{\mathcal{I}_k}\left(\mathbf{z}\right)\right) (\boldsymbol{\Sigma}_{\mathcal{I}_k}^{1/2})^{\mathrm{H}} \mathbf{t}_j.
\end{aligned}
$$

Since γ and α_k's are non-negative and $\boldsymbol{\Sigma}_{\mathcal{I}_k}^{1/2}\mathrm{adj}(\boldsymbol{\Phi}_{\mathcal{I}_k}(\mathbf{z}))(\boldsymbol{\Sigma}_{\mathcal{I}_k}^{1/2})^{\mathrm{H}} \succeq 0$, it holds $\mathcal{J}_{\mathrm{dft}}(\bar{\mathbf{z}}) \geq \mathcal{J}_{\mathrm{dft}}(\mathbf{z})$. In other words, the optimal solution is always achieved on the boundary of $\mathcal{Z}$.

Appendix 6: Proof of Lemma 5.2 and Lemma 5.3

Since Lemma 5.2 can be viewed as a special case of Lemma 5.3, we focus on the proof of the latter. Suppose there exists a feasible solution $\mathbf{x} \in [\mathbf{p}, \mathbf{q}] - [\bar{\mathbf{p}}, \bar{\mathbf{q}}]$ satisfying $\mathcal{E}_{\text{dft}}^{\text{pcp}}(\mathbf{x}) = \mathcal{E}_{\text{dft}}^{+}(\mathbf{x}) - \mathcal{E}_{\text{dft}}^{-}(\mathbf{x}) \geq \mathcal{E}_{\text{dft}}^{\text{cbv}}$. Given that $\mathcal{E}_{\text{dft}}^{+}(\cdot)$ and $\mathcal{E}_{\text{dft}}^{-}(\cdot)$ are monotonically increasing, it can be seen that $\mathcal{E}_{\text{dft}}^{+}(\mathbf{x}) \geq \mathcal{E}_{\text{dft}}^{\text{cbv}} + \mathcal{E}_{\text{dft}}^{-}(\mathbf{p})$ from the chain of inequalities $\mathcal{E}_{\text{dft}}^{+}(\mathbf{q}) \geq \mathcal{E}_{\text{dft}}^{+}(\mathbf{x}) \geq \mathcal{E}_{\text{dft}}^{\text{cbv}} + \mathcal{E}_{\text{dft}}^{-}(\mathbf{x}) \geq \mathcal{E}_{\text{dft}}^{\text{cbv}} + \mathcal{E}_{\text{dft}}^{-}(\mathbf{p})$.

The assumption $\mathbf{x} \notin [\bar{\mathbf{p}}, \bar{\mathbf{q}}]$ translates to either $\mathbf{x} \not\geq \bar{\mathbf{p}}$ or $\mathbf{x} \not\leq \bar{\mathbf{q}}$. Suppose that $\mathbf{x} \not\geq \bar{\mathbf{p}}$, i.e., there exists $j \notin \{i \mid [\mathbf{p}]_i = 0, [\mathbf{q}]_i = 1\}$ such that $[\mathbf{x}]_j = 0$ and $[\bar{\mathbf{p}}]_j = 1$. This implies $\mathbf{x} \leq \mathbf{q} - \mathbf{e}_j$. In view of the conditions for reduction in (5.24), on one hand, if $\mathcal{E}_{\text{dft}}^{+}(\mathbf{q} - \mathbf{e}_j) - \mathcal{E}_{\text{dft}}^{-}(\mathbf{p}) < \mathcal{E}_{\text{dft}}^{\text{cbv}}$, then $\mathcal{E}_{\text{dft}}^{+}(\mathbf{x}) \leq \mathcal{E}_{\text{dft}}^{+}(\mathbf{q} - \mathbf{e}_j) < \mathcal{E}_{\text{dft}}^{\text{cbv}} + \mathcal{E}_{\text{dft}}^{-}(\mathbf{p})$, which however contradicts the assumption that $\mathcal{E}_{\text{dft}}^{+}(\mathbf{x}) \geq \mathcal{E}_{\text{dft}}^{\text{cbv}} + \mathcal{E}_{\text{dft}}^{-}(\mathbf{p})$. On the other hand, if $h_{+}(\mathbf{q} - \mathbf{e}_j) < 0$, then $h_{+}(\mathbf{x}) \leq h_{+}(\mathbf{q} - \mathbf{e}_j) < 0$, contradicting the feasibility of $\mathbf{x}$. Therefore, it must hold $\mathbf{x} \geq \bar{\mathbf{p}}$. The case of $\mathbf{x} \leq \bar{\mathbf{q}}$ can be similarly proved.

References

1. T. Yoo and A. Goldsmith, "On the optimality of multiantenna broadcast scheduling using zero-forcing beamforming," *IEEE J. Sel. Areas Commun.*, vol. 24, no. 3, pp. 528–541, Mar. 2006.
2. C. B. Peel, B. M. Hochwald, and A. L. Swindlehurst, "A vector-perturbation technique for near-capacity multiantenna multiuser communication - Part I: Channel inversion and regularization," *IEEE Trans. Commun.*, vol. 53, no. 1, pp. 195–202, Jan. 2005.
3. B. M. Hochwald, C. B. Peel, and A. L. Swindlehurst, "A vector-perturbation technique for near-capacity multiantenna multiuser communication - Part II: Perturbation," *IEEE Trans. Commun.*, vol. 53, no. 3, pp. 537–544, Mar. 2005.
4. C. B. Chae, S. H. Kim, and R. W. Heath, "Block diagonalized vector perturbation for multiuser MIMO systems," *IEEE Trans. Wireless Commun.*, vol. 7, no. 11, pp. 4051–4057, Nov. 2008.
5. J. Park, B. Lee, and B. Shim, "A MMSE vector precoding with block diagonalization for multiuser MIMO downlink," *IEEE Trans. Commun.*, vol. 60, no. 2, pp. 569–577, Feb. 2012.
6. W. Yao, S. Chen, and L. Hanzo, "A transceiver design based on uniform channel decomposition and MBER vector perturbation," *IEEE Trans. Veh. Technol.*, vol. 59, no. 6, pp. 3153–3159, Jul. 2010.
7. D. Ying, F. W. Vook, T. A. Thomas, D. J. Love, and A. Ghosh, "Kronecker product correlation model and limited feedback codebook design in a 3D channel model," in *Proc. IEEE Int. Conf. Commun.*, Sydney, Australia, Jun. 2014.
8. D.-S. Shiu, G. J. Foschini, M. J. Gans, and J. M. Kahn, "Fading correlation and its effect on the capacity of multielement antenna systems," *IEEE Trans. Commun.*, vol. 48, no. 3, pp. 502–513, Mar. 2000.
9. A. Adhikary, J. Nam, J.-Y. Ahn, and G. Caire, "Joint spatial division and multiplexing - The large-scale array regime," *IEEE Trans. Inf. Theory*, vol. 59, no. 10, pp. 6441–6463, Oct. 2013.
10. D. A. Schmidt, M. Joham, and W. Utschick, "Minimum mean square error vector precoding," *European Trans. Telecommun.*, vol. 19, no. 3, pp. 219–231, Apr. 2008.
11. E. Agrell, T. Eriksson, A. Vardy, and K. Zeger, "Closest point search in lattices," *IEEE Trans. Inf. Theory*, vol. 48, no. 8, pp. 2201–2214, Aug. 2002.
12. X.-W. Chang, "MILES: MATLAB package for solving mixed integer least squares problems," School of Computer Science, McGill University, http://www.cs.mcgill.ca/~chang/software/MILES.php, Jun. 2016.

13. J. H. Conway and N. J. A. Sloane, *Sphere Packings, Lattices and Groups*, ser. Grundlehren der mathematischen Wissenschaften. New York, NY, USA: Springer-Verlag, 2013.
14. P. A. Absil, R. Mahony, and R. Sepulchre, *Optimization Algorithms on Matrix Manifolds*. Princeton, NJ, USA: Princeton University Press, 2007.
15. J. Nocedal and S. J. Wright, *Numerical Optimization*, 2nd ed. New York, NY, USA: Springer, 2006.
16. A. Liu and V. K. N. Lau, "Phase only RF precoding for massive MIMO systems with limited RF chains," *IEEE Trans. Signal Process.*, vol. 62, no. 17, pp. 4505–4515, Sep. 2014.
17. H. Tuy, M. Minoux, and N. T. Hoai-Phuong, "Discrete monotonic optimization with application to a discrete location problem," *SIAM J. Optim.*, vol. 17, no. 1, pp. 78–97, 2006.
18. P. A. Absil, R. Mahony, and J. Trumpf, "An extrinsic look at the Riemannian Hessian," in *Geometric Science of Information*, ser. Lecture Notes in Computer Science, F. Nielsen and F. Barbaresco, Eds. Springer Berlin Heidelberg, 2013, ch. 10, pp. 361–368.
19. CVX Research, Inc., "CVX: Matlab software for disciplined convex programming, version 2.0," http://cvxr.com/cvx, Aug. 2012.
20. A. Adhikary, E. Al Safadi, M. K. Samimi, R. Wang, G. Caire, T. S. Rappaport, and A. F. Molisch, "Joint spatial division and multiplexing for mm-wave channels," *IEEE J. Sel. Areas Commun.*, vol. 32, no. 6, pp. 1239–1255, Jun. 2014.
21. D. Kim, G. Lee, and Y. Sung, "Two-stage beamformer design for massive MIMO downlink by trace quotient formulation," *IEEE Trans. Commun.*, vol. 63, no. 6, pp. 2200–2211, Jun. 2015.
22. J. T. Chen and V. K. N. Lau, "Two-tier precoding for FDD multi-cell massive MIMO time-varying interference networks," *IEEE J. Sel. Areas Commun.*, vol. 32, no. 6, pp. 1230–1238, Jun. 2014.
23. R. A. Horn and C. R. Johnson, *Matrix Analysis*, 2nd ed. New York, NY, USA: Cambridge University Press, 2013.
24. D. J. Ryan, I. B. Collings, I. V. L. Clarkson, and R. W. Heath, "Performance of vector perturbation multiuser MIMO systems with limited feedback," *IEEE Trans. Commun.*, vol. 57, no. 9, pp. 2633–2644, Sep. 2009.
25. J. Choi, "Multiuser precoding with limited cooperation for large-scale MIMO multicell downlink," *IEEE Trans. Wireless Commun.*, vol. 14, no. 3, pp. 1295–1308, Mar. 2015.
26. H. Shin and J. H. Lee, "Capacity of multiple-antenna fading channels: Spatial fading correlation, double scattering, and keyhole," *IEEE Trans. Inf. Theory*, vol. 49, no. 10, pp. 2636–2647, Oct. 2003.
27. A. Edelman, T. Arias, and S. Smith, "The geometry of algorithms with orthogonality constraints," *SIAM J. Matrix Anal. Appl.*, vol. 20, no. 2, pp. 303–353, Oct. 1998.

Chapter 6
Nonlinear Hybrid Precoding for Massive MIMO with Universal Frequency Reuse

6.1 Introduction

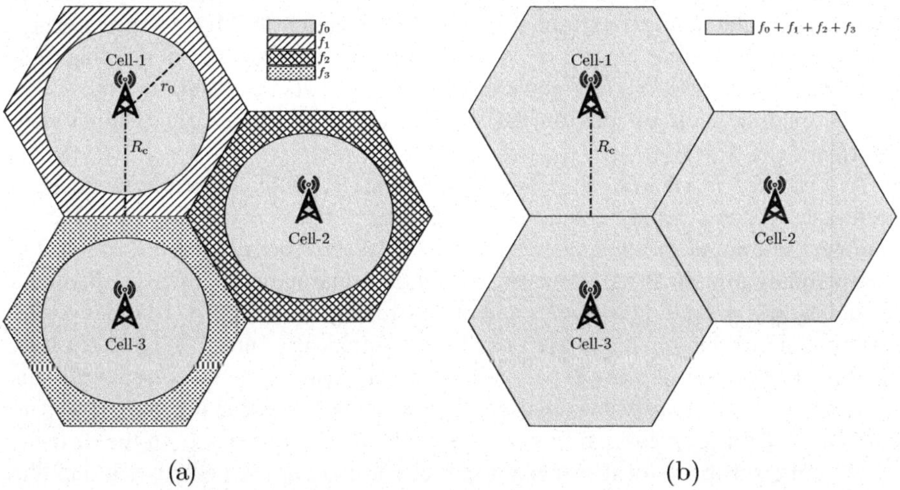

Fig. 6.1 (**a**) Fractional frequency reuse versus (**b**) universal frequency reuse

In an attempt to mitigate the performance degradation caused by inter-cell interference (ICI) in cellular networks with universal frequency reuse, frequency planning techniques, such as fractional frequency reuse, have been proposed [1]. As illustrated in Fig. 6.1a, the basic idea is to partition each cell area into a cell-center region, as indicated by the interior of a circle with a radius r_0, and a cell-edge region. Accordingly, a sub-band represented by a subset of subcarriers at a carrier frequency, say f_0, is universally assigned to the cell-center areas across

T. Le-Ngoc, R. Mai, *Hybrid Massive MIMO Precoding in Cloud-RAN*, Wireless Networks, https://doi.org/10.1007/978-3-030-02158-0_6

the entire cellular network while non-overlapping sub-bands, i.e., orthogonal sets of subcarriers corresponding to different carrier frequencies f_1, f_2, f_3, are separately allocated to each cell-edge area. In doing this, the spatial separation, i.e., path loss, between areas that share the same frequency resources is increased. Through careful optimization of the sub-band assignment and the frequency reuse radius r_0, it is possible to decrease the ICI so much so that it suffices to employ single-cell processing. This is indeed the rationale behind considering nonlinear hybrid precoding in a single-cell environment in Chap. 5.

One drawback of ICI management via frequency planning arises from the difficulty with dynamic optimization in practice. As one has to settle for suboptimal frequency planning schemes [2], the precious frequency resources are not fully exploited. With the increasing popularity of small cells which are characterized by unplanned node placement and irregular infrastructure of deployment [3], the lack of dynamic optimization also brings into question the applicability of such frequency-domain techniques. One promising alternative, known as coordinated multi-point transmission/reception (CoMP) in 3GPP Long-Term Evolution (LTE), is to introduce coordination and cooperation between cells in the spatial domain while allowing universal frequency reuse. In the context of hybrid RF-baseband precoding for the massive multiple-input multiple-output (MIMO) broadcast channel, the RF precoding stage serves as a natural candidate for combating ICI, and lays the foundation for single-cell processing at baseband. In this chapter, we explore such an idea on the assumption of two-timescale channel state information (CSI).

As observed in Chap. 5, the error performance for fully loaded homogeneous multi-user systems could be substantially enhanced by incorporating the effect of nonlinear minimum mean square error (MMSE)-vector perturbation (VP) into the RF precoding optimization. However, it is worth pointing out that the effectiveness of such hybrid precoding solutions and other existing ones, e.g., [4–6] relies on the assumption that the scheduled users can be geographically partitioned into clusters, and the same cluster of users experiences identical channel spatial correlation. This naturally leads one to ask how the user clusters can be formed in the first place. The authors in [7–9] proposed various user grouping algorithms based on the distance between the subspaces of the transmit spatial correlation. For each user group, the RF beamformer was adapted to the mean of the correlation matrices. Since the mean correlation is only a rough approximation, in this case, it is unlikely that the resulting RF beams create near-perfect spatial separation, which casts doubt on the feasibility of cluster-wise spatial multiplexing at baseband. When it comes to a multi-cell environment with universal frequency reuse, one potential issue with deploying such designs originates from the observation that the transmit correlated subspaces of the cell-edge user clusters tend to significantly overlap with each other due to their geographical proximity, which thus renders it ineffective to adjust the RF beamforming solely based on the statistical spatial information. Therefore, a rethink on how to exploit the two-timescale CSI more effectively for hybrid precoding is needed.

In a mobile environment, temporal correlation among consecutive channel fading blocks can generally be observed. We are therefore interested to study if the performance of hybrid precoding with MMSE-VP can benefit from such additional statistical CSI. In particular, two-timescale CSI is assumed to be available at the base stations (BSs), which consists of the perfect knowledge of the slow-varying statistical CSI, i.e., the channel spatial and temporal correlation, and the imperfect instantaneous observations of the RF-beamformed channel. Our focus is on serving clustered users at the cell edge, which indicate the limiting performance of a cellular network. We note that by leveraging the low-dimensional effective CSI as partial observations, the high-dimensional MIMO channel can be estimated and tracked via Kalman filtering. This motivates us to exploit such channel estimates in place of the channel spatial correlation for the RF beamformer design in conjunction with centralized and distributed MMSE-VP at baseband, respectively.

For centralized MMSE-VP, we derive the baseband solution by alternating optimization, which is robust to the observation noise of the RF-beamformed effective channel. Recognizing that the objective function is biconvex, we propose a Jacobi scheme to update its dual formulation with reduced complexity. Although the baseband solution needs to be iteratively obtained, its optimal structure nonetheless reveals that the objective function can be simplified to a quadratic form. Built upon this insight, we then propose a non-iterative approach to joint RF-baseband optimization. Specifically, by abstracting the effect of nonlinear baseband precoding, we arrive at a separate formulation of RF beamforming, and circumvent the complication of RF-baseband alternating optimization. With the objective of balancing the accuracy of channel tracking and overall system mean square error (MSE), the RF beamformer is numerically obtained by a Cayley transformation-based gradient descent search. For distributed MMSE-VP, we propose an approximate block diagonalization (BD) technique for RF beamforming based on the channel estimates from Kalman filtering. Furthermore, an alternative RF beamforming solution is developed in an attempt to relax the restrictive condition of BD beamforming that the number of RF chains be equal to that of users, and to enhance the error performance in the absence of initial channel estimates. The benefits of hybrid precoding with channel tracking are numerically demonstrated in terms of the increased flexibility of generating beamforming directions, and the alleviated sensitivity to channel estimation errors.

The remainder of this chapter is organized as follows. In Sect. 6.2, we introduce the system model and formally state the problem for multi-cell hybrid precoding with MMSE-VP. The strategies of integrating hybrid precoding with channel tracking are developed for centralized MMSE-VP in Sect. 6.3 and for distributed MMSE-VP in Sect. 6.4, respectively. Performance evaluation is numerically conducted in Sect. 6.5, and concluding remarks are provided in Sect. 6.6.

6.2 System Model and Problem Statement

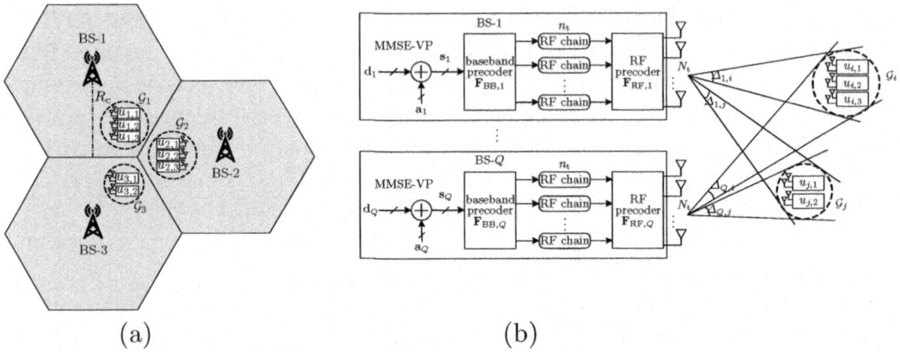

Fig. 6.2 Nonlinear hybrid precoding for multiple clusters of users in a multi-cell massive MIMO downlink with two-timescale CSI. (**a**) Overview. (**b**) General schematic diagram

As illustrated in Fig. 6.2, let us consider a multi-cell massive MIMO system, where a group of base stations (BSs) $\mathcal{Q} \triangleq \{1, \ldots, Q\}$, each equipped with a large-scale antenna array of $N_t \gg 1$ elements, serves K single-antenna users at the cell edge. The transmit antenna array at each BS is driven by n_t RF chains, satisfying $N_t \gg n_t$. Let $\mathcal{K}_q = \{u_{q,1}, \ldots, u_{q,K_q}\}$ denote the K_q users associated with BS-q, and $\mathcal{K}_q \cap \mathcal{K}_l = \emptyset, \forall l \neq q$. We assume a block fading channel with a coherence block of T channel uses. The received signal by user-$u_{q,i}$ reads as

$$y_{q,i}[t] = \sum_{b\in\mathcal{Q}} \mathbf{h}_{b,q,i}^{[n]\mathrm{H}} \mathbf{F}_{\mathrm{RF},b}^{[n]} \mathbf{x}_b[t] + n_{q,i}[t],$$

where $[t]$ denotes the tth channel use of the nth fading block with $(n-1)T < t \leq nT$, $\mathbf{h}_{b,q,i}^{[n]} \in \mathbb{C}^{N_t}$ represents the channel coefficients between BS-b and user-$u_{q,i}$, $\mathbf{F}_{\mathrm{RF},b}^{[n]} \in \mathbb{C}^{N_t \times n_t}$ and $\mathbf{x}_b[t] \in \mathbb{C}^{n_t}$ respectively represent the corresponding RF analog precoder and the baseband-precoded signals at cell-b, and $n_{q,i}[t] \sim \mathcal{CN}(0, \sigma_n^2)$ is the spatially white additive Gaussian noise with variance σ_n^2. Let $\mathbf{H}_{b,q}^{[n]} \triangleq [\mathbf{h}_{b,q,1}^{[n]}, \ldots, \mathbf{h}_{b,q,K_q}^{[n]}]^{\mathrm{H}} \in \mathbb{C}^{K_q \times N_t}$. Collectively for $\mathcal{K}_q$, we can write

$$\mathbf{y}_q[t] = \sum_{b\in\mathcal{Q}} \mathbf{H}_{b,q}^{[n]} \mathbf{F}_{\mathrm{RF},b}^{[n]} \mathbf{x}_b[t] + \mathbf{n}_q[t],$$

where $\mathbf{y}_q[t] \triangleq (y_{q,k}[t])_{k\in\mathcal{K}_q}$, and $\mathbf{n}_q[t] \triangleq (n_{q,k}[t])_{k\in\mathcal{K}_q}$. Specifically, the spatially correlated channel is expressed as $\mathbf{h}_{b,q,i}^{[n]} = \mathbf{\Sigma}_{b,q,i}^{1/2} \mathbf{h}_{b,q,i}^{\mathrm{w}[n]}$, where $\mathbf{\Sigma}_{b,q,i} = \mathbb{E}[\mathbf{h}_{b,q,i}^{[n]} \mathbf{h}_{b,q,i}^{[n]\mathrm{H}}]$ is the transmit correlation matrix, and $\mathbf{h}_{b,q,i}^{\mathrm{w}[n]} \sim \mathcal{CN}(\mathbf{0}, \mathbf{I}_{N_t})$ represents the small-scale Rayleigh fading channel. Through spectral decomposition $\mathbf{\Sigma}_{b,q,i} =$

$\mathbf{U}_{b,q,i}\boldsymbol{\Lambda}_{b,q,i}\mathbf{U}_{b,q,i}^{\mathrm{H}}$, we define $\boldsymbol{\Sigma}_{b,q,i}^{1/2} \triangleq \mathbf{U}_{b,q,i}\boldsymbol{\Lambda}_{b,q,i}^{1/2}$. For the practical case of one-ring local scattering [4, 10], the spatial correlation coefficient between the rth and sth antenna at the BS is generically given by

$$[\boldsymbol{\Sigma}]_{r,s} = \frac{\alpha_{\mathrm{pl}}}{\Delta}\int_{\bar{\theta}-\frac{\Delta}{2}}^{\bar{\theta}+\frac{\Delta}{2}} e^{\jmath 2\pi d_{\lambda}(r-s)\sin(\bar{\theta}-\theta)}d\theta, \tag{6.1}$$

where α_{pl} is the path loss, d_{λ} is the normalized inter-antenna spacing, $\bar{\theta}$ is the mean azimuth angle of departure of scatterers, and Δ is the angle spread of the scattering environment. In a mobile environment, consecutive channel realizations tend to be temporally correlated, which can be characterized by the popular first-order stationary Gauss-Markov model [5], i.e.,

$$\mathbf{h}_{b,q,i}^{[n+1]} = \xi_{q,i}\mathbf{h}_{b,q,i}^{[n]} + \sqrt{1-\xi_{q,i}^{2}}\,\boldsymbol{\Sigma}_{b,q,i}^{1/2}\mathbf{h}_{b,q,i}^{\mathrm{evo}[n]}, \tag{6.2}$$

where $\mathbf{h}_{b,q,i}^{\mathrm{evo}[n]} \sim \mathcal{CN}(\mathbf{0}, \mathbf{I}_{N_{\mathrm{t}}})$ represents the evolution across successive fading blocks, and $\xi_{q,i}$ is the temporal fading coefficient. It is assumed that the perfect knowledge of long-term channel statistics, as represented by $\{\boldsymbol{\Sigma}_{b,q,i}\}$ and $\{\xi_{q,i}\}$, is available at the BSs, which can be estimated during the period of uplink channel training [11]. For notational simplicity, we omit the index $[n]$ of the nth channel coherence block unless otherwise stated, and denote $\mathcal{T}_n \triangleq \{t' \in \mathbb{Z} | (n-1)T < t' \leq nT\}$ as the corresponding set of channel use indices.

For BS-q, the RF analog precoder $\mathbf{F}_{\mathrm{RF},q}$ is assumed to satisfy orthonormality to avoid strong intra-cell interference, i.e., $\mathbf{F}_{\mathrm{RF},q}^{\mathrm{H}}\mathbf{F}_{\mathrm{RF},q} = \mathbf{I}_{n_{\mathrm{t}}}$. For analytical simplicity, the sample average power constraint over a fading block of T channel uses is imposed, i.e., $\frac{1}{T}\sum_{t\in\mathcal{T}_n}\|\mathbf{x}_q[t]\|^2 \leq P_{\mathrm{t}}, q \in \mathcal{Q}$, where the same power constraint is assumed across all BSs without loss of generality (w.l.o.g.). The data streams destined for the users in $\mathcal{K}_q$, denoted as $\mathbf{d}_q[t] \triangleq (d_{q,k}[t])_{k\in\mathcal{K}_q} \in \mathbb{C}^{K_q}$, are first perturbed by a complex integer vector $\mathbf{a}_q[t] \triangleq (a_{q,k}[t])_{k\in\mathcal{K}_q} \in \mathbb{Z}^{K_q} + \jmath\mathbb{Z}^{K_q}$, and then linearly precoded by $\mathbf{F}_{\mathrm{BB},q} \in \mathbb{C}^{n_{\mathrm{t}}\times K_q}$ to generate $\mathbf{x}_q[t]$. In particular, the perturbed signal $\mathbf{s}_q[t] \triangleq (s_{q,k}[t])_{k\in\mathcal{K}_q}$ reads as

$$\mathbf{s}_q[t] = \mathbf{d}_q[t] + \tau\mathbf{a}_q[t],$$

and the digitally precoded signal at baseband is accordingly given as $\mathbf{x}_q[t] = \mathbf{F}_{\mathrm{BB},q}\mathbf{s}_q[t]$. Here, the design parameter $\tau > 0$ is chosen to ensure correct signal recovery by modulo decoding and control the effect of the perturbation vector [12]. With the element-wise modulo-τ operation $f_{\tau}(\mathbf{z}) = \mathbf{z} - \left\lfloor \frac{\mathbf{z}}{\tau} + 0.5 \right\rfloor \tau$, the perturbation vector $\mathbf{a}_q[t]$ can be removed as

$$\mathbf{d}_q[t] = f_{\tau}(\Re\{\mathbf{s}_q[t]\}) + \jmath f_{\tau}(\Im\{\mathbf{s}_q[t]\}).$$

At the output of the scalar equalizer $\beta \in \mathbb{C}$, the signal estimates $\hat{\mathbf{s}}_q[t] \triangleq (\hat{s}_{q,k}[t])_{k\in\mathcal{K}_q}$ for $\mathcal{K}_q$ are given as

$$\hat{\mathbf{s}}_q[t] = \bar{\beta}\mathbf{y}_q[t] = \bar{\beta}\sum_{b\in\mathcal{Q}} \mathbf{H}_{b,q}\mathbf{F}_{\mathrm{RF},b}\mathbf{x}_b[t] + \bar{\beta}\mathbf{n}_q[t],$$

and the MSE of cell-q conditioned on $\{\mathbf{d}_q[t]\}$ is defined as

$$\begin{aligned}\mathcal{E}_q[t] &\triangleq \mathbb{E}_{\mathbf{n}_q}\left[\|\mathbf{s}_q[t] - \hat{\mathbf{s}}_q[t]\|^2 \big| \{\mathbf{d}_q[t]\}\right] \\ &= |\beta|^2\left(\textstyle\sum_{i,j\in\mathcal{Q}}\mathbf{x}_j[t]^{\mathrm{H}}\mathbf{G}_{j,q}^{\mathrm{H}}\mathbf{G}_{i,q}\mathbf{x}_i[t] + K_q\sigma_n^2\right) \\ &\quad - 2\Re\left\{\beta\textstyle\sum_{j\in\mathcal{Q}}\mathbf{x}_j[t]^{\mathrm{H}}\mathbf{G}_{j,q}^{\mathrm{H}}\mathbf{s}_q[t]\right\} + \|\mathbf{s}_q[t]\|^2. \end{aligned} \tag{6.3}$$

where $\mathbf{G}_{q,b} \triangleq \mathbf{H}_{q,b}\mathbf{F}_{\mathrm{RF},q} \in \mathbb{C}^{K_b\times n_\mathrm{t}}$ for $q, b \in \mathcal{Q}$ denotes the low-dimensional effective MIMO channel seen at baseband. Note that the MSE for baseband VP is defined as the squared distance between the perturbed signals and the signal estimates before modulo decoding. The conditional MSE is motivated by the observation that the statistics of the perturbed signal are generally not known. Accordingly, the nonlinear hybrid precoder design is formulated as a solution for minimizing the sum MSE across a transmission block of T channel uses, i.e.,

$$\begin{aligned}&\min \sum_{q\in\mathcal{Q}}\sum_{t\in\mathcal{T}_n} \mathcal{E}_q[t] \\ &\text{s.t. } \mathbf{F}_{\mathrm{RF},q}^{\mathrm{H}}\mathbf{F}_{\mathrm{RF},q} = \mathbf{I}_{n_\mathrm{t}}, \sum_{t\in\mathcal{T}_n}\|\mathbf{x}_q[t]\|^2 \le TP_\mathrm{t}, q \in \mathcal{Q},\end{aligned}$$

with variables $\{\mathbf{F}_{\mathrm{RF},q}, \mathbf{x}_q[t], \mathbf{a}_q[t], \beta\}_{q\in\mathcal{Q},t\in\mathcal{T}_n}$. To facilitate such a two-timescale CSI-based hybrid design, in what follows, we first derive the nonlinear MMSE-VP precoder at baseband, which reduces the objective function to a simplified form. In view of the difficulty with solving for the RF stage due to its dependence on the baseband, the simplification allows us to abstract the baseband processing, and hence obviates the need for complex RF-baseband alternating optimization. This is motivated by the desire to reduce channel training overhead, where the receivers have available only the local knowledge of the RF-precoded channels. On the contrary, since users associated with the same BS employ the same scalar combiner, it can be centrally computed and made known to the users by cell-wide broadcasting. Thus, we jointly address scalar combining with baseband precoding with respect to each channel use. In Sect. 6.3, in the context of centralized MMSE-VP, we study the design of hybrid precoding. In an attempt to reduce computational complexity and the requirement on CSI exchange, local CSI-based distributed updating is addressed in Sect. 6.4.

6.3 Hybrid Precoder Design with Centralized MMSE-VP

In this section, we first derive iteratively the digital MMSE-VP solution at baseband. To decouple the RF design from the baseband, the effect of nonlinear baseband processing is abstracted. The RF beamformer is then designed based on the channel estimates produced by Kalman filtering.

6.3.1 Robust Centralized Baseband Precoding

Suppose that imperfect observations of the RF-precoded MIMO channels $\mathbf{G}_{q,b} = \mathbf{H}_{q,b}\mathbf{F}_{\mathrm{RF},q}$ for $q, b \in \mathcal{Q}$ are available at baseband, and are modeled as

$$\mathbf{G}_{q,b} = \bar{\mathbf{G}}_{q,b} + \Omega_{q,b},$$

where the spatially white channel estimation error $\Omega_{q,b} \in \mathbb{C}^{K_b \times n_\mathrm{t}}$ is distributed as $\mathrm{vec}(\Omega_{q,b}) \sim \mathcal{CN}(\mathbf{0}, \sigma_e^2 \mathbf{I}_{n_\mathrm{t}} \otimes \mathbf{I}_{K_b})$ with variance σ_e^2 determined by the chosen channel estimation scheme. To obtain a robust design of baseband precoders and equalizers that accounts for channel estimation errors, the following optimization problem is formulated, i.e.,

$$\begin{aligned} &\min \sum_{q\in\mathcal{Q}} \sum_{t\in\mathcal{T}_n} \mathbb{E}_{\Omega_{q,b}}[\mathcal{E}_q[t]] \\ &\text{s.t.} \sum_{t\in\mathcal{T}_n} \|\mathbf{x}_q[t]\|^2 \le T P_\mathrm{t}, q \in \mathcal{Q}, \end{aligned} \tag{6.4}$$

with variables $\{\mathbf{x}_q[t], \beta, \mathbf{a}_q[t]\}_{q\in\mathcal{Q}, t\in\mathcal{T}_n}$. Let $\bar{\mathcal{E}}_q[t] \triangleq \mathbb{E}_{\Omega_{q,b}}[\mathcal{E}_q[t]]$ and $\bar{\mathcal{E}} = \sum_{q\in\mathcal{Q}} \sum_{t\in\mathcal{T}_n} \bar{\mathcal{E}}_q[t]$. For the purpose of jointly deriving the baseband solutions across all BSs, the objective function $\bar{\mathcal{E}}_q[t]$ is evaluated and rewritten as

$$\begin{aligned} \bar{\mathcal{E}}_q[t] = {} & |\beta|^2 \left[\mathbf{x}[t]^\mathrm{H} \left(\bar{\mathbb{G}}_q^\mathrm{H} \bar{\mathbb{G}}_q + K_q \sigma_e^2 \mathbf{I}_{Qn_\mathrm{t}}\right) \mathbf{x}[t] + K_q \sigma_n^2\right] \\ & - 2\Re\left\{\beta \mathbf{x}[t]^\mathrm{H} \bar{\mathbb{G}}_q^\mathrm{H} \mathbf{s}_q[t]\right\} + \left\|\mathbf{s}_q[t]\right\|^2, \end{aligned} \tag{6.5}$$

where $\bar{\mathbb{G}}_q$ represents the noisy observation of $\mathbb{G}_q \triangleq \mathbb{H}_q \mathbf{F}_\mathrm{RF}$ with $\mathbb{H}_q \triangleq [\mathbf{H}_{1,q}, \ldots, \mathbf{H}_{Q,q}]$, $\mathbf{F}_\mathrm{RF} \triangleq \mathrm{diag}\{\mathbf{F}_{\mathrm{RF},1}, \ldots, \mathbf{F}_{\mathrm{RF},Q}\}$, and the precoded signals at baseband are aggregately written as $\mathbf{x}[t] \triangleq (\mathbf{x}_q[t])_{q\in\mathcal{Q}}$. Furthermore, let $\mathbf{s}[t] \triangleq (\mathbf{s}_q[t])_{q\in\mathcal{Q}}$, $\hat{\mathbf{s}}_q[t] \triangleq (\hat{\mathbf{s}}_q[t])_{q\in\mathcal{Q}}$, and $\mathbf{n}[t] \triangleq (\mathbf{n}_q[t])_{q\in\mathcal{Q}}$ represent the aggregate perturbed signals, the signal estimates, and the noise across all cells, respectively. We define the Lagrangian as

$$\mathcal{L}(\mathbf{x}_q[t], \beta, \boldsymbol{\lambda}) = \bar{\mathcal{E}} + \sum_{q\in\mathcal{Q}} \lambda_q \left(\sum_{t\in\mathcal{T}_n} \|\mathbf{x}_q[t]\|^2 - T P_{\mathrm{t}} \right), \tag{6.6}$$

where $\boldsymbol{\lambda} \triangleq (\lambda_q)_{q\in\mathcal{Q}} \succeq \mathbf{0}$ is the Lagrangian multipliers. Let $\bar{\mathbb{G}} \triangleq [\bar{\mathbb{G}}_1^{\mathrm{T}}, \ldots, \bar{\mathbb{G}}_Q^{\mathrm{T}}]^{\mathrm{T}}$. Setting the first-order derivative of the Lagrangian to zero, i.e., $\frac{\partial \mathcal{L}}{\partial \mathbf{x}[t]^{\mathrm{H}}} = \mathbf{0}$ and $\frac{\partial \mathcal{L}}{\partial \beta^*} = 0$, we have

$$\mathbf{x}^{\star}[t] = \beta^{\star} \left(|\beta^{\star}|^2 \bar{\mathbb{G}}^{\mathrm{H}} \bar{\mathbb{G}} + |\beta^{\star}|^2 K \sigma_e^2 \mathbf{I}_{Qn_{\mathrm{t}}} + \boldsymbol{\Lambda} \right)^{-1} \bar{\mathbb{G}}^{\mathrm{H}} \mathbf{s}[t], \tag{6.7}$$

where $\mathbf{x}^{\star}[t]$ is seen to have a regularized zero-forcing (RZF) precoder with $\boldsymbol{\Lambda} = \boldsymbol{\lambda} \otimes \mathbf{I}_{n_{\mathrm{t}}}$ as the linear front end, and the scalar equalizer is given by

$$\beta^{\star} = \frac{\sum_{t\in\mathcal{T}_n} \mathbf{s}[t]^{\mathrm{H}} \bar{\mathbb{G}} \mathbf{x}[t]}{\sum_{t\in\mathcal{T}_n} \mathbf{x}[t]^{\mathrm{H}} \left(\bar{\mathbb{G}}^{\mathrm{H}} \bar{\mathbb{G}} + K\sigma_e^2 \mathbf{I} \right) \mathbf{x}[t] + T K \sigma_n^2}. \tag{6.8}$$

It can be seen that $\bar{\mathcal{E}}$ is a biconvex function, i.e., $\bar{\mathcal{E}}$ is convex with respect to $(\mathbf{x}[t])_{t\in\mathcal{T}_n}$ (β) when β ($(\mathbf{x}[t])_{t\in\mathcal{T}_n}$) is held fixed. Naturally, a limit point can be generated through a Gauss-Seidel update rule which iterates between solving (6.4) for $(\mathbf{x}[t])_{t\in\mathcal{T}_n}$ with β given, and updating β according to (6.8) until convergence is reached. Because the objective function is lower bounded by zero and monotonically decreased at each iteration, convergence of the algorithm is guaranteed. Since the feasible region is closed and convex, the resulting limit point is a stationary point [13]. Note that the dimension of the primal variables $(\mathbf{x}[t])_{t\in\mathcal{T}_n} \in \mathbb{C}^{TQn_{\mathrm{t}}}$ is generally not small. Fortunately, conditioned on β, the quadratically constrained quadratic program (QCQP) of $(\mathbf{x}[t])_{t\in\mathcal{T}_n}$ (6.4) admits strong duality, and hence allows us to solve the equivalent dual problem of $\boldsymbol{\lambda} \in \mathbb{C}^{Q}$, which has a smaller dimension as a result of the limited number of cooperating BSs in practice. Once $\boldsymbol{\lambda}$ becomes available, the baseband precoded solution $(\mathbf{x}[t])_{t\in\mathcal{T}_n}$ can be constructed from (6.7).

To complete the baseband design, the perturbation vector needs to be determined. Substituting in (6.7) and (6.8), the objective function can be reduced as

$$\mathcal{E} = \sum_{t\in\mathcal{T}_n} \mathbf{s}[t]^{\mathrm{H}} \underbrace{\left[\mathbf{I}_K + \bar{\mathbb{G}} (K\sigma_e^2 \mathbf{I} + |\beta^{\star}|^{-2} \boldsymbol{\Lambda})^{-1} \bar{\mathbb{G}}^{\mathrm{H}} \right]^{-1}}_{\triangleq \boldsymbol{\Phi}} \mathbf{s}[t], \tag{6.9}$$

which follows directly from the matrix inversion lemma. We observe that it suffices to consider a real scalar equalizer $\beta \in \mathbb{R}$. Since $\boldsymbol{\Phi}$ is positive definite (PD), by Cholesky factorization we write $\boldsymbol{\Phi} = \mathbf{L}^{\mathrm{H}}\mathbf{L}$, where $\mathbf{L}$ is upper triangular with positive diagonal elements. Given (6.9), the search for the perturbation vectors can be formulated as T integer least squares (ILS) problems, i.e., for $t \in \mathcal{T}_n$,

$$\begin{aligned}\mathbf{a}^{\star}[t] &= \underset{\mathbf{a}[t]\in\mathbb{Z}^K+\jmath\mathbb{Z}^K}{\arg\min}\ \mathbf{s}[t]^{\mathrm{H}}\boldsymbol{\Phi}\mathbf{s}[t] \\ &= \underset{\mathbf{a}[t]\in\mathbb{Z}^K+\jmath\mathbb{Z}^K}{\arg\min}\ \|-\mathbf{L}\mathbf{d}[t]-\tau\mathbf{L}\mathbf{a}[t]\|^2. \end{aligned} \tag{6.10}$$

Although such a problem is NP-hard, through Lenstra-Lenstra-Lovász (LLL)-reduction-based preprocessing, it can be efficiently solved by the Schnorr-Euchner enumeration strategy when $\mathbf{L}$ is full rank [14]. It is seen that $\mathbf{L}$ depends on β and $\boldsymbol{\lambda}$. Naturally, one might be tempted to consider an iterative procedure that optimizes the linear baseband solutions $\{\beta, \boldsymbol{\lambda}\}$ and the perturbation vectors $(\mathbf{a}[t])_{t\in\mathcal{T}_n}$ alternatively. Unfortunately, there is no guarantee that such a procedure would lead to a monotonically decreasing objective $\mathcal{E}$. This is because the iterates $\{\beta, \boldsymbol{\lambda}, (\mathbf{a}[t])_{t\in\mathcal{T}_n}\}$ are not necessarily feasible in the absence of the power constraint in (6.10). On the other hand, to ensure the feasibility of the iterates, a joint search of perturbation vectors across T channel uses is required. This is however computationally prohibitive with $T \gg 1$. In view of these challenges, we heuristically set $\beta^{-2}\boldsymbol{\Lambda} = \frac{K\sigma_n^2}{QP}\mathbf{I}_{Qn_{\mathrm{t}}}$ in $\boldsymbol{\Phi}$, which is the optimal regularizing coefficient corresponding to the sum power constraint [15]. We will see that such approximation allows us to abstract the overall effect of VP over T channel uses, and facilitates the RF beamformer design in the next section. Once the perturbation vectors are found, the linear baseband solutions $\{\beta, \boldsymbol{\lambda}\}$ are successively optimized so that the per-BS power constraint is satisfied.

We note that in the derivation of the baseband solutions $(\mathbf{x}[t])_{t\in\mathcal{T}_n}$, specifically the RZF precoder $\mathbf{F}_{\mathrm{BB}}$ from the dual variable $\boldsymbol{\lambda}$ at each iteration, direct matrix inversion needs to be carried out, where the computational complexity scales on the order of $\mathcal{O}(Q^3 n_{\mathrm{t}}^3)$ [16]. This might quickly become expensive as Q and n_{t} increase. To address this issue, one alternative is to compute $(\mathbf{x}_q[t])_{t\in\mathcal{T}_n}, q \in \mathcal{Q}$ in parallel instead of $(\mathbf{x}[t])_{t\in\mathcal{T}_n}$ as a whole. Towards this end, by setting the first-order derivative of $\mathcal{L}$ in (6.6) with respect to $\mathbf{x}_q^{\mathrm{H}}[t]$ to zero, i.e., $\frac{\partial\mathcal{L}}{\partial\mathbf{x}_q^{\mathrm{H}}[t]} = \mathbf{0}$, we have

$$\mathbf{x}_q^{\star}[t] = \beta^{-1}\boldsymbol{\Phi}_{q,q}^{-1}\bar{\mathbf{G}}_{q,q}^{\mathrm{H}}\mathbf{s}_q[t] - \boldsymbol{\eta}_q^{\star}[t], \tag{6.11}$$

where $\boldsymbol{\Phi}_{q,q} \triangleq \sum_{b\in\mathcal{Q}}\bar{\mathbf{G}}_{q,b}^{\mathrm{H}}\bar{\mathbf{G}}_{q,b} + (K\sigma_e^2 + \frac{\lambda_q}{\beta^2})\mathbf{I}_{n_{\mathrm{t}}}$, and the term $\boldsymbol{\eta}_q[t]$ representing the residual ICI reads as

$$\boldsymbol{\eta}_q^{\star}[t] \triangleq \boldsymbol{\Phi}_{q,q}^{-1}\sum_{b\in\mathcal{Q}\backslash q}\left(\bar{\mathbf{R}}_{q,b}\mathbf{x}_b^{\star}[t] - \beta^{-1}\bar{\mathbf{G}}_{q,b}^{\mathrm{H}}\mathbf{s}_b[t]\right),$$

with $\bar{\mathbf{R}}_{q,b} \triangleq \sum_{l\in\mathcal{Q}}\bar{\mathbf{G}}_{q,l}^{\mathrm{H}}\bar{\mathbf{G}}_{b,l} + \delta_{q,b}K\sigma_e^2\mathbf{I}_{n_{\mathrm{t}}}$. In doing this, the dimension of the matrix to be inverted per iteration is decreased from Qn_{t} to n_{t}. To further simplify the computation, we write $\bar{\mathbf{R}}_{q,q} = \bar{\mathbf{U}}_{q,q}\bar{\boldsymbol{\Sigma}}_{q,q}\bar{\mathbf{U}}_{q,q}^{\mathrm{H}}$ by the eigenvalue decomposition, where $\bar{\mathbf{U}}_{q,q}$ is unitary and $\bar{\boldsymbol{\Sigma}}_{q,q}$ is diagonal. Consequently,

$$\boldsymbol{\Phi}_{q,q}^{-1} = (\bar{\mathbf{R}}_{q,q} + \lambda_q \beta^{-2}\mathbf{I}_{n_t})^{-1} = \bar{\mathbf{U}}_{q,q}(\bar{\boldsymbol{\Sigma}}_{q,q} + \lambda_q \beta^{-2}\mathbf{I}_{n_t})^{-1}\bar{\mathbf{U}}_{q,q}^{\mathrm{H}}.$$

Hence, the complexity of updating the matrix inverse is reduced to $\mathcal{O}(1)$ per iteration once the eigenvalue decomposition of $\bar{\mathbf{R}}_{q,q}$ is made available at the initialization stage.

Let us take the update of $\mathbf{x}_q[t]$ for example, and denote $\mathbf{x}_{-q}[t] \triangleq (\mathbf{x}_b[t])_{b\in\mathcal{Q}\setminus q}$. We observe that when $\{\mathbf{x}_{-q}[t]\}_{t\in\mathcal{T}_n}$ and β are fixed as the current, say the νth, iterate, the original nonconvex optimization problem becomes a strongly convex QCQP. Following the Jacobi algorithm, a solution to such a convex program provides a search direction for the next iterate. In particular, for $t \in \mathcal{T}_n$, the next iterate is obtained as

$$\mathbf{x}_q^{\nu+1}[t] = \mathbf{x}_q^{\nu}[t] + \gamma^{\nu}\big(\hat{\mathbf{x}}_q^{\nu}[t] - \mathbf{x}_q^{\nu}[t]\big), q \in \mathcal{Q}.$$

In computing the search direction $\boldsymbol{\delta}_q^{\nu}[t] \triangleq \hat{\mathbf{x}}_q^{\nu}[t] - \mathbf{x}_q^{\nu}[t]$ for $t \in \mathcal{T}_n$, $\{\hat{\mathbf{x}}_q^{\nu}[t]\}_{t\in\mathcal{T}_n}$ is posed as a solution to

$$\begin{aligned} &\min\ \bar{\mathcal{E}}_q^{\nu}(\{\mathbf{x}_q[t]\}_{t\in\mathcal{T}_n}) \\ &\text{s.t.} \sum_{t\in\mathcal{T}_n} \|\mathbf{x}_q[t]\|^2 \le T P_{\mathrm{t}}, \end{aligned} \tag{6.12}$$

where $\{\mathbf{x}_q[t]\}_{t\in\mathcal{T}_n}$ are the variables, and the objective function $\bar{\mathcal{E}}_q^{\nu}(\{\mathbf{x}_q[t]\}_{t\in\mathcal{T}_n})$ is obtained by replacing $\{\mathbf{x}_{-q}[t]\}_{t\in\mathcal{T}_n}$ and β in (6.5) with the current iterates $\{\mathbf{x}_{-q}^{\nu}[t]\}_{t\in\mathcal{T}_n}$ and β^{ν}, respectively. Besides, since the objective function is quadratic, the step length γ^{ν} can be determined by simply projecting the solution $\gamma_{\mathrm{sol}}^{\nu}$ of the linear equation

$$\frac{d}{d\gamma^{\nu}}\mathcal{E}(\{\mathbf{x}_q^{\nu+1}[t]\}_{t\in\mathcal{T}_n}, \{(\mathbf{x}_{-q}^{\nu}[t]\}_{t\in\mathcal{T}_n}, \beta^{\nu}) = 0,$$

i.e.,

$$\gamma_{\mathrm{sol}}^{\nu} = \frac{\Re\big\{\beta \sum_{t\in\mathcal{T}_n} \boldsymbol{\delta}^{\nu}[t]^{\mathrm{H}}\bar{\mathbb{G}}^{\mathrm{H}}\left(\mathbf{s}[t] - \beta^{*}\bar{\mathbb{G}}\mathbf{x}^{\nu}[t]\right)\big\}}{|\beta|^2 \sum_{t\in\mathcal{T}_n} \boldsymbol{\delta}^{\nu}[t]^{\mathrm{H}}(\bar{\mathbb{G}}^{\mathrm{H}}\bar{\mathbb{G}} + K\sigma_e^2\mathbf{I})\boldsymbol{\delta}^{\nu}[t]}, \tag{6.13}$$

onto $[0, 1]$, where $\boldsymbol{\delta}^{\nu}[t] \triangleq (\boldsymbol{\delta}_q^{\nu}[t])_{q\in\mathcal{Q}}$. Furthermore, the scalar equalizer is updated as

$$\beta^{\nu+1} = \frac{\sum_{t\in\mathcal{T}_n} \mathbf{s}[t]^{\mathrm{H}}\bar{\mathbb{G}}\mathbf{x}^{\nu+1}[t]}{\sum_{t\in\mathcal{T}_n} \mathbf{x}^{\nu+1}[t]^{\mathrm{H}}\left(\bar{\mathbb{G}}^{\mathrm{H}}\bar{\mathbb{G}} + K\sigma_e^2\mathbf{I}\right)\mathbf{x}^{\nu+1}[t] + TK\sigma_n^2}. \tag{6.14}$$

It is not difficult to see that the dual problem to (6.12) is scalar, whose equivalence is guaranteed by the convexity of the primal problem. Let us define the Lagrangian with respect to the qth cell as

$$\mathcal{L}_q^{\nu}(\{\mathbf{x}_q[t]\}) \triangleq \bar{\mathcal{E}}_q^{\nu}(\{\mathbf{x}_q[t]\}) + \mu_q \sum_{t\in\mathcal{T}_n} \|\mathbf{x}_q[t]\|^2,$$

Algorithm 6.1 Baseband precoding solution by Jacobi algorithm

Output: Stationary solutions $\{\mathbf{x}_q^{\eta}[t]\}_{q\in\mathcal{Q},t\in\mathcal{T}_n}$.
1: **Initialization:** Set $\eta = 0$, $\gamma^{\eta} = 1$, $(\mathbf{x}_q^{\eta}[t])_{q,t}$, and β^{η}.
2: **while** true **do**
3: For all $q \in \mathcal{Q}$, compute $\{\hat{\mathbf{x}}_q^{\eta}[t]\}_{t\in\mathcal{T}_n}$ by solving (6.12) through its dual (6.15);
4: If $\frac{1}{T}\sum_{t\in\mathcal{T}_n} \|\hat{\mathbf{x}}_q^{\eta}[t] - \mathbf{x}_q^{\eta}[t]\|_{\infty} < \epsilon$, break; otherwise, update $\mathbf{x}_q^{\eta+1}[t]$ as

$$\mathbf{x}_q^{\eta+1}[t] \leftarrow \mathbf{x}_q^{\eta}[t] + \gamma^{\eta}\left(\hat{\mathbf{x}}_q[t] - \mathbf{x}_q^{\eta}[t]\right), \forall q, t$$

with the step length γ^{η} determined by (6.13);
5: Set $\eta \leftarrow \eta + 1$;
6: **end while**

Algorithm 6.2 Linear baseband solution by Gauss-Seidel update

Output: Stationary solutions $\{\mathbf{x}_q^{\nu}[t]\}_{q\in\mathcal{Q},t\in\mathcal{T}_n}$, β.
1: **Initialization:** Set $\nu = 0$, $\gamma^{\nu} = 1$, $\{\mathbf{x}_q^{\nu}[t]\}_{q,t}$, and β^{ν}.
2: **while** true **do**
3: Compute $\{\mathbf{x}_q^{\nu}[t]\}_{q,t}$ by Algorithm 6.1;
4: Update $\beta^{\nu+1}$ based on (6.14) by setting $\mathbf{x}_q[t] = \mathbf{x}_q^{\nu+1}[t]$;
5: If $\frac{1}{T}\sum_{t\in\mathcal{T}_n} \|\hat{\mathbf{x}}_q^{\nu}[t] - \mathbf{x}_q^{\nu}[t]\|_{\infty} < \epsilon$, break;
6: Set $\nu \leftarrow \nu + 1$;
7: **end while**

and substitute in (6.11) where the current iterate $\{\mathbf{x}_b^{\nu}[t]\}_{b\in\mathcal{Q}\backslash q}$ is used in $\boldsymbol{\eta}_q[t]$. The optimal dual variable can be found by solving the nonlinear equation $d\mathcal{L}_q^{\nu}/d\lambda_q = 0$, i.e.,

$$\sum_{t\in\mathcal{T}_n} \left[\boldsymbol{\eta}_q^{\nu}[t]^{\mathrm{H}}\boldsymbol{\Phi}_{q,q}^{-1}\left(\mathbf{I}_{n_t} - 2\boldsymbol{\Phi}_{q,q}^{-1}\boldsymbol{\Psi}_{q,q}\right)\boldsymbol{\Phi}_{q,q}^{-1}\boldsymbol{\eta}_q^{\nu}[t] - 2\Re\left\{\tilde{\mathbf{s}}_q[t]^{\mathrm{H}}\bar{\mathbf{G}}_{q,q}\boldsymbol{\Phi}_{q,q}^{-2}\boldsymbol{\eta}_q^{\nu}[t]\right\}\right] = 0, \tag{6.15}$$

where $\tilde{\mathbf{s}}_q[t] \triangleq \beta\sum_{b\in\mathcal{Q}\backslash q} \mathbf{G}_{b,q}\mathbf{x}_b^{\nu}[t] - \mathbf{s}_q[t]$, and $\boldsymbol{\Psi}_{q,q} \triangleq \bar{\mathbf{G}}_{q,q}^{\mathrm{H}}\bar{\mathbf{G}}_{q,q} + (K_q\sigma_e^2 + \lambda_q\beta^{-2})\mathbf{I}_{n_t}$. The details are formally stated in Algorithms 6.1 and 6.2.

Remark 6.1 Convergence of the Jacobi algorithm follows directly from the observations that (1) conditioned on β and $\{\mathbf{x}_{-q}[t]\}_{t\in\mathcal{T}_n}$, the objective function is strongly convex in $\{\mathbf{x}_q[t]\}_{t\in\mathcal{T}_n}$, and (2) the feasible region dictated by the power constraint is compact and convex [17].

6.3.2 Joint RF-Baseband Precoding

Let $\mathcal{E}^\star(\{\mathbf{F}_{\mathrm{RF},q}\}_{q\in\mathcal{Q}})$ denote the resulting MSE from the baseband solutions $\{\beta^\star,\boldsymbol{\lambda}^\star,(\mathbf{a}^\star[t])_{t\in\mathcal{T}_n}\}$, which we recall reads as $\mathcal{E}^\star(\{\mathbf{F}_{\mathrm{RF},q}\}) = \sum_{t\in\mathcal{T}_n}\mathbf{s}^\star[t]^{\mathrm{H}}\boldsymbol{\Phi}^\star\mathbf{s}^\star[t]$ with $\boldsymbol{\Phi}^\star$ obtained by substituting in $\boldsymbol{\lambda}^\star$. Design of the RF components of the hybrid precoders is accordingly formulated as

$$\begin{aligned} &\min \ \mathcal{E}^\star\left(\{\mathbf{F}_{\mathrm{RF},q}\}_{q\in\mathcal{Q}}\right) \\ &\text{s.t.} \ \mathbf{F}_{\mathrm{RF},q}^{\mathrm{H}}\mathbf{F}_{\mathrm{RF},q} = \mathbf{I}_{n_\mathrm{t}},\ q\in\mathcal{Q} \end{aligned} \tag{6.16}$$

with variables $\{\mathbf{F}_{\mathrm{RF},q}\}_{q\in\mathcal{Q}}$. The fact that the baseband solutions $\{\beta,\boldsymbol{\lambda}\}$ are iteratively updated implies that the baseband variables cannot be eliminated from $\mathcal{E}^\star(\{\mathbf{F}_{\mathrm{RF},q}\}_{q\in\mathcal{Q}})$. In other words, optimization of $\mathcal{E}^\star(\{\mathbf{F}_{\mathrm{RF},q}\}_{q\in\mathcal{Q}})$ with respect to $\{\mathbf{F}_{\mathrm{RF},q}\}_{q\in\mathcal{Q}}$ depends on the availability of the baseband solutions, which in turn relies on the noisy observation of the effective channel $\{\mathbf{G}_{q,b}\}_{q,b\in\mathcal{Q}}$ by assumption. Such interdependence, coupled with the nonlinear search for the perturbation vectors, makes it difficult to find an optimal solution to (6.16). Instead, we propose to abstract the nonlinear baseband processing, where the linear baseband components $\{\beta,\boldsymbol{\lambda}\}$ subject to the per-BS power constraint are approximated by their sum power-constrained counterpart, i.e., $\beta^{-2}\boldsymbol{\Lambda} = \frac{K\sigma_n^2}{QP}\mathbf{I}_{Qn_\mathrm{t}}$. On the other hand, to approximate the perturbation effect over an entire block of T channel uses, we note that when T is asymptotically large, it holds

$$\lim_{T\to\infty}\frac{\mathcal{E}(\{\mathbf{F}_{\mathrm{RF},q}\}_q)}{T} = \mathbb{E}_{\mathbf{s}^\star}\left[\mathbf{s}^{\star\mathrm{H}}\boldsymbol{\Phi}\mathbf{s}^\star\right] = \mathbb{E}_{\mathbf{s}^\star}\left[\|\mathbf{L}\mathbf{s}^\star\|^2\right],$$

where the ensemble average is taken with respect to $\mathbf{s}^\star = \mathbf{s}^\star[t]$ at an arbitrary time instant. Generally, the expectation cannot be exactly evaluated to a closed form in that the statistics of the perturbed signals $(\mathbf{s}^\star[t])_{t\in\mathcal{T}_n}$ are not obtainable. Fortunately, by treating the signal constellation as continuously and symmetrically distributed around the origin, $\mathbf{L}\mathbf{s}^\star$ is uniformly distributed as a result of the optimal perturbation vector search [18]. This is a reasonable relaxation when high-order square QAM modulation, e.g., 16-, 64-, 256-QAM, is employed [19]. Following a similar line of derivation as in [19], a lower bound for the stochastic objective function can be established as

$$\mathbb{E}_{\mathbf{s}_q}\left[\mathcal{E}^{\star}(\{\mathbf{F}_{\mathrm{RF},q}\})\right] \geq \underbrace{\frac{QK\tau^2}{\pi e}\det(\boldsymbol{\Phi})^{\frac{1}{QK}}}_{\triangleq \mathcal{E}_{\mathrm{vp}}(\{\mathbf{F}_{\mathrm{RF},q}\})}.$$

For the sake of mathematical tractability, we instead optimize the deterministic lower bound $\mathcal{E}_{\mathrm{vp}}(\{\mathbf{F}_{\mathrm{RF},q}\})$. However, such an approximate problem is still nontrivial since both the objective and the feasible region are non-convex. The lack of direct access to full CSI at RF makes the issue even more complicated.

Let us consider the nth channel fading block. Recall that by assumption the reduced-dimensional observations of the high-dimensional MIMO channel are available at the BSs, as represented by

$$\mathbf{g}_{b,q,i}^{[n]} \triangleq \mathbf{F}_{\mathrm{RF},b}^{[n]\mathrm{H}}\mathbf{h}_{b,q,i}^{[n]} + \boldsymbol{\omega}_{b,q,i}^{[n]}, \forall b, q, i,$$

where we recall that $\mathbf{h}_{b,q,i}^{[n]}$ is the channel vector between BS-b and user-$u_{q,i}$ (the ith user associated with the qth cell). In view of the linear state-space characterization of the channel in (6.2), we propose to exploit the partially observed CSI for estimation of the full MIMO channel. In fact, this can be formulated as a linear system estimation problem, and solved in the sense of MMSE by Kalman filtering. Specifically, the Kalman filtering update rules relate the channel estimate $\hat{\mathbf{h}}_{b,q,i}^{n|n}$, the prediction error covariance matrix $\mathbf{P}_{b,q,i}^{n|n-1}$, and the estimation error covariance matrix $\mathbf{P}_{b,q,i}^{n|n}$ iteratively as

$$\hat{\mathbf{h}}_{b,q,i}^{n|n} = \hat{\mathbf{h}}_{b,q,i}^{n|n-1} + \mathbf{K}_{b,q,i}^{[n]}(\mathbf{g}_{b,q,i}^{[n]} - \mathbf{F}_{\mathrm{RF},b}^{[n]\mathrm{H}}\hat{\mathbf{h}}_{b,q,i}^{n|n-1}), \tag{6.17}$$

$$\mathbf{P}_{b,q,i}^{n|n-1} = \xi_{q,i}^2\mathbf{P}_{b,q,i}^{n-1|n-1} + \left(1 - \xi_{q,i}^2\right)\boldsymbol{\Sigma}_{b,q,i}, \tag{6.18}$$

$$\mathbf{P}_{b,q,i}^{n|n} = \mathbf{P}_{b,q,i}^{n|n-1} - \mathbf{K}_{b,q,i}^{[n]}\mathbf{F}_{\mathrm{RF},b}^{[n]\mathrm{H}}\mathbf{P}_{b,q,i}^{n|n-1}, \tag{6.19}$$

where $\mathbf{K}_{b,q,i}^{[n]} = \mathbf{P}_{b,q,i}^{n|n-1}\mathbf{F}_{\mathrm{RF},b}^{[n]}(\mathbf{F}_{\mathrm{RF},b}^{[n]\mathrm{H}}\mathbf{P}_{b,q,i}^{n|n-1}\mathbf{F}_{\mathrm{RF},b}^{[n]} + \sigma_e^2\mathbf{I}_{n_t})^{-1}$ is the gain matrix, and the update starts from the initial channel prediction $\hat{\mathbf{h}}_{b,q,i}^{1|0}$ and the covariance matrix $\mathbf{P}_{b,q,i}^{1|0}$. In particular, $\hat{\mathbf{h}}_{b,q,i}^{1|0} = \mathbf{0}$ and $\mathbf{P}_{b,q,i}^{1|0} = \boldsymbol{\Sigma}_{b,q,i}$ in the absence of initial channel estimates. On the other hand, if imperfect channel estimates are obtainable, as modeled by

$$\mathbf{h}_{b,q,i}^{[0]} = \tilde{\mathbf{h}}_{b,q,i}^{[0]} + \mathbf{e}_{b,q,i}^{[0]},$$

where $\mathbf{e}_{b,q,i}^{[0]} \sim \mathcal{CN}(\mathbf{0}, \sigma_e^2 \mathbf{I}_{N_t})$ with variance σ_e^2 represents the channel estimation errors, the predicted channel is initialized as $\hat{\mathbf{h}}_{b,q,i}^{1|0} = \tilde{\mathbf{h}}_{b,q,i}^{[0]}$, and the corresponding prediction error covariance is directly calculated as

$$\mathbf{P}_{b,q,i}^{1|0} = \hat{\mathbf{h}}_{b,q,i}^{1|0} \left(\hat{\mathbf{h}}_{b,q,i}^{1|0}\right)^{\mathrm{H}} + \sigma_e^2 \mathbf{I}_{N_t}.$$

It is seen that in addition to the sum MSE, the RF beamformers $\{\mathbf{F}_{\mathrm{RF},q}^{[n]}\}$ also have an impact on the accuracy of channel tracking as represented by the error covariance matrices $\{\mathbf{P}_{b,q,i}^{n|n}\}$, which in turn affects the effectiveness of beam separation in the RF domain. Let us define the normalized channel estimation errors as

$$\mathcal{E}_{\mathrm{kf}}^{[n]} \triangleq \sum_{b \in \mathcal{Q}} \sum_{i \in \mathcal{K}_b} \mathrm{tr}\left(\mathbf{P}_{b,b,i}^{n|n}\right). \tag{6.20}$$

Interestingly, when users within the same cluster experience the same channel statistics as considered in [6, 20], the RF precoder that achieves optimal channel tracking is directly related to the eigenvectors of the prediction covariance matrix.

Theorem 6.1 *Suppose that users from the same cluster share the identical channel spatial and temporal correlation. The optimal RF precoders in the sense of minimizing the channel tracking errors* $\mathcal{E}_{\mathrm{kf}}^{[n]}$ *are given by the eigenvectors of* $(\mathbf{P}_{q,q,i}^{n|n-1} + \sigma_e^2 \mathbf{I}_{n_t})^{-1} (\mathbf{P}_{q,q,i}^{n|n-1})^2$ *corresponding to the* n_t *largest eigenvalues.*

Proof See Appendix 1.

Since the system error performance is the ultimate objective, we propose to design the RF precoders by balancing the VP-related MSE and the channel tracking accuracy, which is formulated as

$$\begin{aligned} &\min \ \mathcal{E}_{\mathrm{vp}}^{[n]} + \omega_{\mathrm{kf}} \mathcal{E}_{\mathrm{kf}}^{[n]} \\ &\text{s.t. } \mathbf{F}_{\mathrm{RF},q}^{[n]\mathrm{H}} \mathbf{F}_{\mathrm{RF},q}^{[n]} = \mathbf{I}_{n_t}, \, q \in \mathcal{Q}, \end{aligned}$$

with variables $\{\mathbf{F}_{\mathrm{RF},q}^{[n]}\}_{q \in \mathcal{Q}}$ and the weight $\omega_{\mathrm{kf}} > 0$. Note that the VP-related MSE $\mathcal{E}_{\mathrm{vp}}^{[n]}$ for the nth block is optimized by treating the estimated channel $\hat{\mathbf{h}}_{b,q,i}^{n|n}$, $\forall b, q, i$, as the true instantaneous CSI. On the other hand, derivation of such channel estimates in (6.17) relies on the noisy observations of the effective channels $\{\mathbf{g}_{b,q,i}^{[n]}\}$, which unfortunately depend on the design variables $\{\mathbf{F}_{\mathrm{RF},q}^{[n]}\}$. Here, we circumvent this difficulty by using the channel prediction $\hat{\mathbf{h}}_{b,q,i}^{n|n-1}$. Accordingly, the VP-related approximate MSE is modified as

$$\hat{\mathcal{E}}_{\mathrm{vp}}^{[n]}(\{\mathbf{F}_{\mathrm{RF},q}\}) \triangleq \frac{QK\tau^2}{\pi e} \det\left(\hat{\boldsymbol{\Phi}}^{n|n-1}\right)^{\frac{1}{QK}},$$

where $\hat{\boldsymbol{\Phi}}^{n|n-1}$ is defined by replacing the actual channel realization $\mathbb{H}$ in $\boldsymbol{\Phi}$ with the channel estimates $\{\hat{\mathbf{h}}_{b,q,i}^{n|n-1}\}$. Once the RF beamformers become available, the observations of the current fading block $\{\mathbf{g}_{b,q,i}^{[n]}\}$ can be obtained through downlink channel training and feedback, and the channel estimate is updated according to (6.17), (6.18), and (6.19).

For notational convenience, we omit the channel use index $[n]$ and the superscript $\cdot^{n|n-1}$ when no confusion can be caused. Let $\hat{\mathcal{E}} \triangleq \hat{\mathcal{E}}_{\text{vp}} + \omega_{\text{kf}}\mathcal{E}_{\text{kf}}$ denote the objective function, which is nonconvex but differentiable. This allows us to find a stationary point by gradient descent-like algorithms. However, direct application of algorithms developed for the Euclidean space is not suitable since the iterates generated are generally not orthonormal and thus not feasible. Here, we adapt the framework from [21], where the feasibility of the updates is guaranteed by the Cayley transformation-based parameterization, and the parameter is optimized through a curvilinear search. Let us take for example the update of $\mathbf{F}_{\text{RF},b}$ w.l.o.g. In that the orthonormal constraints are not coupled with each other, the update rule applies equally to all the RF precoding matrices. Let $\mathbf{F}_{\text{RF},b}^{\nu}$ denote the current feasible iterate. Then the update is parameterized by α through the Cayley transformation as

$$\mathbf{F}_{\text{RF},b}^{\nu}(\alpha) = \left(\mathbf{I}_{N_\text{t}} + \frac{\alpha}{2}\boldsymbol{\Omega}_{\mathbf{F}_{\text{RF},b}^{\nu}}\right)^{-1}\left(\mathbf{I}_{N_\text{t}} - \frac{\alpha}{2}\boldsymbol{\Omega}_{\mathbf{F}_{\text{RF},b}^{\nu}}\right)\mathbf{F}_{\text{RF},b}^{\nu}, \tag{6.21}$$

which can be verified to satisfy $\mathbf{F}_{\text{RF},b}^{\nu}(\alpha)^{\text{H}}\mathbf{F}_{\text{RF},b}^{\nu}(\alpha) = \mathbf{I}_{n_\text{t}}$ for all $\alpha \in \mathbb{R}$ as long as $\boldsymbol{\Omega}_{\mathbf{F}_{\text{RF},b}^{\nu}}$ is skew-symmetric, i.e., $\boldsymbol{\Omega}_{\mathbf{F}_{\text{RF},b}^{\nu}}^{\text{H}} = -\boldsymbol{\Omega}_{\mathbf{F}_{\text{RF},b}^{\nu}}$ [21]. In this case, the optimization problem of matrix variables is reduced to one of the scalar parameter α at each iteration. One choice of a skew-symmetric $\boldsymbol{\Omega}_{\mathbf{F}_{\text{RF},b}}$ is given by

$$\boldsymbol{\Omega}_{\mathbf{F}_{\text{RF},b}} = [\mathcal{P}_{\mathbf{F}_{\text{RF},b}}(\nabla_{\mathbf{F}_{\text{RF},b}}\hat{\mathcal{E}})]\mathbf{F}_{\text{RF},b}^{\text{H}} - \mathbf{F}_{\text{RF},b}[\mathcal{P}_{\mathbf{F}_{\text{RF},b}}(\nabla_{\mathbf{F}_{\text{RF},b}}\hat{\mathcal{E}})]^{\text{H}},$$

where $\mathcal{P}_{\mathbf{F}_{\text{RF},b}} \triangleq \mathbf{I}_{N_\text{t}} - \frac{1}{2}\mathbf{F}_{\text{RF},b}\mathbf{F}_{\text{RF},b}^{\text{H}}$ defines a projection into the nullspace of $\mathbf{F}_{\text{RF},b}$, and $\nabla_{\mathbf{F}_{\text{RF},b}}\hat{\mathcal{E}} = \nabla_{\mathbf{F}_{\text{RF},b}}\hat{\mathcal{E}}_{\text{vp}} + \nabla_{\mathbf{F}_{\text{RF},b}}\mathcal{E}_{\text{kf}}$ denotes the Euclidean gradient at $\mathbf{F}_{\text{RF},b}$. Although matrix inversion is involved in (6.21), the numerical computation can be greatly simplified by exploiting the structure of $\boldsymbol{\Omega}_{\mathbf{F}_{\text{RF},b}}$ together with the fact that the dimensions of the design variables satisfy $N_\text{t} \gg n_\text{t}$ [21]. It is not difficult to show that

$$\nabla_{\mathbf{F}_{\text{RF},b}}\hat{\mathcal{E}}_{\text{vp}} = -\frac{2\tau^2}{\pi e}\det(\hat{\boldsymbol{\Phi}})^{\frac{1}{QK}}\mathbf{S}_{\mathbf{F}_{\text{RF},b}}^{r\text{T}}\hat{\mathbb{H}}^{\text{H}}\hat{\mathbb{H}}\mathbf{F}\hat{\boldsymbol{\Xi}}\mathbf{S}_{\mathbf{F}_{\text{RF},b}}^{c}, \tag{6.22}$$

where we define $\hat{\boldsymbol{\Xi}} \triangleq (\mathbf{F}_{\text{RF}}^{\text{H}}\hat{\mathbb{H}}^{\text{H}}\hat{\mathbb{H}}\mathbf{F}_{\text{RF}} + \beta^{-2}\boldsymbol{\Lambda})^{-1}$ with $\hat{\mathbb{H}}$ denoting the MMSE estimate of $\mathbb{H}$, $\hat{\boldsymbol{\Phi}} = \mathbf{I}_{QK} - \hat{\mathbb{H}}\mathbf{F}_{\text{RF}}\hat{\boldsymbol{\Xi}}\mathbf{F}_{\text{RF}}^{\text{H}}\hat{\mathbb{H}}^{\text{H}}$, and the selection matrices $\mathbf{S}_{\mathbf{F}_{\text{RF},b}}^{r}$ and $\mathbf{S}_{\mathbf{F}_{\text{RF},b}}^{c}$ are defined through $\mathbf{S}_{\mathbf{F}_{\text{RF},b}}^{r\text{T}}\mathbf{F}_{\text{RF}}\mathbf{S}_{\mathbf{F}_{\text{RF},b}}^{c} = \mathbf{F}_{\text{RF},b}$. Furthermore,

$$\nabla_{\mathbf{F}_{\mathrm{RF},b}} \mathcal{E}_{\mathrm{kf}} = 2 \sum_{q \in \mathcal{Q}} \sum_{i \in \mathcal{K}_q} \mathbf{P}_{b,q,i}^{n|n-1} (\mathbf{F}_{\mathrm{RF},b} \mathbf{K}_{b,q,i}^{\mathrm{H}} - \mathbf{I}_{N_{\mathrm{t}}}) \mathbf{K}_{b,q,i}. \tag{6.23}$$

From (6.22) and (6.23), we easily construct the gradient of $\hat{\mathcal{E}}$ with respect to $\mathbf{F}_{\mathrm{RF},b}$ as

$$\nabla_{\mathbf{F}_{\mathrm{RF},b}} \hat{\mathcal{E}} = \nabla_{\mathbf{F}_{\mathrm{RF},b}} \hat{\mathcal{E}}_{\mathrm{vp}} + \omega_{\mathrm{kf}} \nabla_{\mathbf{F}_{\mathrm{RF},b}} \mathcal{E}_{\mathrm{kf}}.$$

Note that the design variables $\{\mathbf{F}_{\mathrm{RF},q}\}$ are parameterized by a single common α. In the cases as such where it is nontrivial to find a globally optimal α, an inexact line search based on Barzilai-Borwein (BB) step size can be used to achieve global convergence of gradient descent methods [21]. The proposed algorithm for the RF beamformer optimization is formally listed in Algorithm 6.3.

Algorithm 6.3 RF beamformer design via gradient descent with BB steps

1: **Initialization:** Set $\nu = 0$ and $\kappa^{\nu} = 1$. Initialize $\mathbf{F}_{\mathrm{RF}}^{\nu} \triangleq \mathrm{blkdiag}\{\mathbf{F}_{\mathrm{RF},1}^{\nu}, \ldots, \mathbf{F}_{\mathrm{RF},Q}^{\nu}\}$ and $0 < \alpha_{\mathrm{lb}} \leq \alpha \leq \alpha_{\mathrm{ub}}$, $0 < \rho, \delta, \eta, \epsilon < 1$. Set $c^{\nu} = \hat{\mathcal{E}}(\mathbf{F}_{\mathrm{RF}}^{\nu})$.
2: **while** $\|\nabla \hat{\mathcal{E}}(\{\mathbf{F}_{\mathrm{RF},q}^{\nu}\})\|_{\mathrm{F}} > \epsilon$ **do**
3:     Generate $\{\boldsymbol{\Omega}_{\mathbf{F}_{\mathrm{RF},q}}^{\nu}\}$ according to (6.22) and (6.23);
4:     **while** $\hat{\mathcal{E}}(\mathbf{F}_{\mathrm{RF}}^{\nu}(\alpha)) \geq c^{\nu} + \rho\alpha\nabla\hat{\mathcal{E}}(\mathbf{F}_{\mathrm{RF}}^{\nu}(0))$ **do**
5:         Set $\alpha \leftarrow \delta\alpha$;
6:     **end while**
7:     For all $q \in \mathcal{Q}$, set $\mathbf{F}_{\mathrm{RF},q}^{\nu+1} = \mathbf{F}_{\mathrm{RF},q}^{\nu}(\alpha)$ and $c^{\nu+1} \leftarrow \frac{\eta\kappa^{\nu}c^{\nu} + \hat{\mathcal{E}}(\{\mathbf{F}_{\mathrm{RF},q}^{\nu+1}\})}{\eta\kappa^{\nu}+1}$;
8:     Set $\alpha \leftarrow \max\{\min(\alpha_{\mathrm{alt}}^{\nu+1}, \alpha_{\mathrm{ub}}), \alpha_{\mathrm{lb}}\}$;
9:     Update $\nu \leftarrow \nu + 1$;
10: **end while**

In Line 4, the gradient term is evaluated as

$$\nabla\hat{\mathcal{E}}\left(\mathbf{F}_{\mathrm{RF}}^{\nu}(0)\right) \triangleq \frac{d}{d\alpha}\hat{\mathcal{E}}\left(\mathbf{F}_{\mathrm{RF}}^{\nu}(\alpha)\right)\Big|_{\alpha=0} = -\frac{1}{2}\|\boldsymbol{\Omega}\|_{\mathrm{F}}^{2},$$

where $\boldsymbol{\Omega} \triangleq \mathrm{blkdiag}\{\boldsymbol{\Omega}_{\mathbf{F}_{\mathrm{RF},1}}, \ldots, \boldsymbol{\Omega}_{\mathbf{F}_{\mathrm{RF},Q}}\}$. In Line 8, $\alpha_{\mathrm{alt}}^{\nu+1}$ is chosen alternatively as $\alpha_1^{\nu+1}$ for an odd ν and as $\alpha_2^{\nu+1}$ for an even ν, where

$$\alpha_1^{\nu+1} \triangleq \frac{\|\mathbf{S}^{\nu}\|_{\mathrm{F}}^{2}}{\left|\mathrm{tr}\left((\mathbf{S}^{\nu})^{\mathrm{H}}\mathbf{Y}^{\nu}\right)\right|}, \alpha_2^{\nu} \triangleq \frac{\left|\mathrm{tr}\left((\mathbf{S}^{\nu})^{\mathrm{H}}\mathbf{Y}^{\nu}\right)\right|}{\|\mathbf{Y}^{\nu}\|_{\mathrm{F}}^{2}},$$

with $\mathbf{S}^{\nu} \triangleq \mathbf{F}_{\mathrm{RF}}^{\nu+1} - \mathbf{F}_{\mathrm{RF}}^{\nu}$ and $\mathbf{Y}^{\nu} \triangleq \nabla\hat{\mathcal{E}}(\mathbf{F}_{\mathrm{RF}}^{\nu+1}) - \nabla\hat{\mathcal{E}}(\mathbf{F}_{\mathrm{RF}}^{\nu})$.

To recap, towards obtaining the hybrid RF-baseband precoder, we first derive the RF solutions $\{\mathbf{F}_{\mathrm{RF},q}\}_{q \in \mathcal{Q}}$ with respect to $\beta^{-2}\boldsymbol{\Lambda} = \frac{K\sigma_n^2}{QP}\mathbf{I}_{Qn_{\mathrm{t}}}$ by Algorithm 6.3. Then through channel training, the instantaneous information of the RF-precoded MIMO channel is made available at baseband, based on which the nonlinear solutions $\{\beta, \boldsymbol{\lambda}, \{\mathbf{a}[t]\}_{t \in \mathcal{T}_n}\}$ at baseband are found.

6.4 Hybrid Precoder Design with Distributed MMSE-VP

6.4.1 Approximate BD-Based RF Beamforming

Although centralized MMSE-VP delivers the best error performance, it incurs backhaul signaling overhead such as exchange of local CSI. Furthermore, ILS search for the perturbation vector, which has worse-case exponential complexity, needs to be jointly performed across all the K users. For applications where the backhaul links cannot afford timely CSI exchange or each cell is heavily loaded with users, it makes sense to consider MMSE-VP in a distributed manner. Unfortunately, in the presence of ICI, as represented by the coupling between the baseband solutions $\mathbf{x}_i$ and $\mathbf{x}_j$ for $i \neq j$, the centralized MMSE-VP formulation in (6.3) does not admit a decomposable structure.

Given the hybrid precoding architecture, it is natural to ask if effective separation can be created in the RF beam domain, where it becomes feasible to focus baseband precoding on each cell. Suppose that as a result of channel tracking as discussed in the previous section, imperfect local CSI estimates $\{\hat{\mathbf{H}}_{q,b}\}_{b\in\mathcal{Q}}$ are available at cell-q. We formulate ICI minimization as

$$\hat{\mathbf{H}}_{q,b}\mathbf{F}_{\mathrm{RF},q} = \mathbf{O}, \forall b \neq q. \tag{6.24}$$

To derive such an RF solution, in the same vein as the development in [22], we form

$$\hat{\boldsymbol{\Theta}}_{-q} \triangleq \left[\hat{\mathbf{H}}_{q,1}^{\mathrm{T}}, \ldots, \hat{\mathbf{H}}_{q,q-1}^{\mathrm{T}}, \hat{\mathbf{H}}_{q,q+1}^{\mathrm{T}}, \ldots, \hat{\mathbf{H}}_{q,Q}^{\mathrm{T}}\right]^{\mathrm{T}},$$

and let the columns of $\hat{\mathbf{B}}_q$ consist of the right singular vectors spanning the null space of $\hat{\boldsymbol{\Theta}}_{-q}$. Furthermore, by SVD, we write $\hat{\mathbf{H}}_{q,q}\hat{\mathbf{B}}_q = \hat{\mathbf{U}}_q\hat{\boldsymbol{\Sigma}}_q\hat{\mathbf{V}}_q^{\mathrm{H}}$, where the singular values on the diagonal of $\hat{\boldsymbol{\Sigma}}_q$ are in non-increasing order, and the columns of $\hat{\mathbf{V}}_q = [\hat{\mathbf{V}}_q^{(1)}, \hat{\mathbf{V}}_q^{(0)}]$ are the right singular vectors with $\hat{\mathbf{V}}_q^{(1)}$ and $\hat{\mathbf{V}}_q^{(0)}$ associated with the nonzero and zero singular values, respectively. Thus, the RF solution satisfying (6.24) reads as

$$\mathbf{F}_{\mathrm{RF},q}^{\mathrm{BD}} = \hat{\mathbf{B}}_q\hat{\mathbf{V}}_q^{(1)}.$$

If the channel estimates $\{\hat{\mathbf{H}}_{q,b}\}$ are sufficiently close to the true CSI $\{\mathbf{H}_{q,b}\}$ in the sense that $\mathbf{H}_{q,b}\mathbf{F}_{\mathrm{RF},q}^{\mathrm{BD}} \approx \mathbf{O}, \forall b \neq q$, the conditional MSE for the qth cell in (6.3) can be approximately reduced to

$$\begin{aligned}\mathcal{E}_q^{\mathrm{BD}}[t] = &\, |\beta_q|^2 \left(\mathbf{x}_q[t]^{\mathrm{H}}\mathbf{F}_{\mathrm{RF},q}^{\mathrm{H}}\mathbf{H}_{q,q}^{\mathrm{H}}\mathbf{H}_{q,q}\mathbf{F}_{\mathrm{RF},q}\mathbf{x}_q[t] + K_q\sigma_n^2\right) \\ &- 2\Re\left\{\beta_q\mathbf{x}_q[t]^{\mathrm{H}}\mathbf{F}_{\mathrm{RF},q}^{\mathrm{H}}\mathbf{H}_{q,q}^{\mathrm{H}}\mathbf{s}_q[t]\right\} + \|\mathbf{s}_q[t]\|^2,\end{aligned}$$

where the problem of multi-cell baseband MMSE-VP is decomposed into Q single-cell problems. Note that in this case, the scalar equalizers $\{\beta_q\}$ are chosen separately for each cell.

6.4.2 Robust Distributed Baseband Precoding

Because the local CSI estimates $\{\hat{\mathbf{H}}_{q,b}\}_{b\in\mathcal{Q}}$ are not perfect, it is unlikely that the condition of complete ICI nulling, i.e., $\mathbf{H}_{q,b}\mathbf{F}^{\mathrm{BD}}_{\mathrm{RF},q} = \mathbf{O}$, $\forall b \in \mathcal{Q}\backslash q$, holds exactly for the true channel realizations $\{\mathbf{H}_{q,b}\}_{b\in\mathcal{Q}\backslash q}$. In an attempt to incorporate into the baseband precoding design the effect of residual ICI and the presence of observation noise in the effective CSI, we redefine the objective function as

$$\begin{aligned}\mathcal{J}_q[t] &\triangleq \mathbb{E}_{\Omega_{q,b}}\left[\mathcal{E}_q^{\mathrm{BD}}[t] + \textstyle\sum_{b\in\mathcal{Q}\backslash q}\|\bar{\beta}_q\mathbf{G}_{q,b}\mathbf{x}_q[t]\|^2\right]\\ &= |\beta_q|^2\left(\mathbf{x}_q[t]^{\mathrm{H}}\bar{\mathbf{R}}_{q,q}\mathbf{x}_q[t] + K_q\sigma_n^2\right) - 2\Re\{\beta_q\mathbf{x}_q[t]^{\mathrm{H}}\bar{\mathbf{G}}_{q,q}^{\mathrm{H}}\mathbf{s}_q[t]\} + \|\mathbf{s}_q[t]\|^2,\end{aligned} \tag{6.25}$$

where we recall that $\bar{\mathbf{R}}_{q,q} = \sum_{b\in\mathcal{Q}}\bar{\mathbf{G}}_{q,b}^{\mathrm{H}}\bar{\mathbf{G}}_{q,b} + K\sigma_e^2\mathbf{I}_{n_{\mathrm{t}}}$. Therefore, the problem for nonlinear baseband precoding with distributed MMSE-VP is stated as

$$\begin{aligned}\min\ &\sum_{t\in\mathcal{T}_n}\sum_{q\in\mathcal{Q}}\mathcal{J}_q[t]\\ \text{s.t.}\ &\sum_{t\in\mathcal{T}_n}\|\mathbf{x}_q[t]\|^2 \le TP_{\mathrm{t}},\end{aligned} \tag{6.26}$$

with variables $\{\mathbf{x}_q[t], \mathbf{a}_q[t], \beta_q\}_{q\in\mathcal{Q},t\in\mathcal{T}_n}$. Let $\mathcal{J}$ denote the objective function of (6.26). Clearly, optimization of the modified sum MSE can be carried out in parallel with respect to each cell. By solving the KKT conditions, the baseband solution for cell-q can be shown as

$$\mathbf{x}_q^\star[t] = \beta_q^{\star-1}\mathbf{W}_q\mathbf{s}_q[t], \tag{6.27}$$

where the linear RZF front end $\mathbf{W}_q \triangleq (\bar{\mathbf{R}}_{q,q} + \xi_q\mathbf{I}_{n_{\mathrm{t}}})^{-1}\bar{\mathbf{G}}_{q,q}^{\mathrm{H}}$ with the regularizing coefficient $\xi_q \triangleq \frac{K_q\sigma_n^2}{P_{\mathrm{t}}}$ depends only on the local effective CSI $\{\bar{\mathbf{G}}_{q,b}\}_{b\in\mathcal{Q}}$ at cell-q, and the optimal $\beta_q^\star$ is chosen to achieve full-power transmission, i.e., $\sum_{t\in\mathcal{T}_n}\|\mathbf{x}_q^\star[t]\|^2 = TP_{\mathrm{t}}$. Indeed, the solution (6.11) is reduced to the one (6.27) with vanishing ICI, i.e., $\|\boldsymbol{\eta}_q\| = 0$. This provides an intuitive justification of the modified MSE in (6.25). We remark that in the absence of information exchange, interference coordination between cells such as power optimization cannot be realized.

By substituting in the baseband solution (6.27), the sum MSE $\mathcal{J}$ in (6.26), can be simplified as

$$\mathcal{J} = \sum_{t\in\mathcal{T}_n}\sum_{q\in\mathcal{Q}} \mathbf{s}_q[t]^{\mathrm{H}}\big[\mathbf{I}_{K_q} - \bar{\mathbf{G}}_{q,q}(\bar{\mathbf{R}}_{q,q} + \xi_q\mathbf{I}_{n_t})^{-1}\bar{\mathbf{G}}_{q,q}^{\mathrm{H}}\big]\mathbf{s}_q[t]$$

$$= \sum_{t\in\mathcal{T}_n}\sum_{q\in\mathcal{Q}} \mathbf{s}_q[t]^{\mathrm{H}}\underbrace{\big[\mathbf{I}_{K_q} + \bar{\mathbf{G}}_{q,q}(\bar{\mathbf{R}}_{q,q}^{-} + \xi_q\mathbf{I}_{n_t})^{-1}\bar{\mathbf{G}}_{q,q}^{\mathrm{H}}\big]^{-1}}_{\triangleq\,\Xi_q}\mathbf{s}_q[t],$$

where $\bar{\mathbf{R}}_{q,q}^{-} \triangleq \bar{\mathbf{R}}_{q,q} - \bar{\mathbf{G}}_{q,q}^{\mathrm{H}}\bar{\mathbf{G}}_{q,q}$, and the second step results from the matrix inversion lemma. By construction, the objective $\mathcal{J}$ admits an additively separable structure and thus allows a parallel search for the perturbation vectors $\{\mathbf{a}_q[t]\}_{q\in\mathcal{Q}}$. By Cholesky factorization $\Xi_q = \mathbf{L}_q^{\mathrm{H}}\mathbf{L}_q$, where $\mathbf{L}_q$ is lower triangular with positive diagonal elements, the search for the perturbation vectors $\{\mathbf{a}_q[t]\}_{q\in\mathcal{Q}}$ is formulated as

$$\mathbf{a}_q^{\star}[t] = \underset{\mathbf{a}_q[t]\in\mathbb{Z}^{K_q}+\jmath\mathbb{Z}^{K_q}}{\arg\min}\ \mathcal{J}_q[t]$$

$$= \underset{\mathbf{a}_q[t]\in\mathbb{Z}^{K_q}+\jmath\mathbb{Z}^{K_q}}{\arg\min}\ \big\|-\mathbf{L}_q\mathbf{d}_q[t] - \tau\mathbf{L}_q\mathbf{a}_q[t]\big\|^2.$$

In contrast with the case of centralized MMSE-VP (6.10), the dimension of the lattice search space for the perturbation vector is decreased from K to K_q. In other words, we trade the flexibility of adjusting the digital precoding solution at baseband for the alleviation of channel estimation overhead and computational complexity.

6.4.3 Joint RF-Baseband Precoding

For the BD-based RF beamformer in Sect. 6.4.1 to exist, the derivation suggests that the same number of users needs to be scheduled as that of RF chains. This condition may somewhat restrict the flexibility of baseband precoding. Following the observation in Sect. 6.4.3, the objective function of the sum MSE for cell-q can be similarly lower bounded as

$$\mathcal{J}_q \geq \mathcal{J}_{\mathrm{vp},q} \triangleq \frac{K_q\tau^2}{\pi e}\det\big(\Xi_q\big)^{\frac{1}{K_q}}, \tag{6.28}$$

which serves as a cost function for the optimization of RF beamforming. From the definition of Ξ_q, it can be seen that the lower bound $\mathcal{J}_{\mathrm{vp},q}$ depends on the RF beamformer implicitly through the averaged noisy observations of the effective channels $\{\bar{\mathbf{G}}_{q,b}\}_{b\in\mathcal{Q}}$. To facilitate the derivation of RF solution, we modify the lower bound as $\hat{\mathcal{J}}_{\mathrm{vp},q} \triangleq \frac{K_q\tau^2}{\pi e}\det\big(\hat{\Xi}_q\big)^{1/K_q}$, where

$$\hat{\boldsymbol{\Xi}}_q \triangleq \left[\mathbf{I}_{K_q} + \hat{\mathbf{H}}_{q,q}\mathbf{F}_{\mathrm{RF},q}\big(\mathbf{F}_{\mathrm{RF},q}^{\mathrm{H}}\hat{\mathbf{R}}_{-q}\mathbf{F}_{\mathrm{RF},q}\big)^{-1}\mathbf{F}_{\mathrm{RF},q}^{\mathrm{H}}\hat{\mathbf{H}}_{q,q}^{\mathrm{H}}\right]^{-1}$$

with $\hat{\mathbf{R}}_{-q} \triangleq \sum_{b\in\mathcal{Q}\backslash q}\hat{\mathbf{H}}_{q,b}^{\mathrm{H}}\hat{\mathbf{H}}_{q,b} + \xi_q\mathbf{I}_{n_t}$, and we recall that $\{\hat{\mathbf{H}}_{q,b}\}_{b\in\mathcal{Q}}$ represent the channel predictions for the current coherence block from Kalman filtering. Interestingly, such an approximation enables us to derive the optimal structure of the RF beamformer with respect to $\hat{\mathcal{J}}_{\mathrm{vp},q}$.

Theorem 6.2 *An optimal RF beamformer $\mathbf{F}_{\mathrm{RF},q}^{\star} \in \mathbb{C}^{N_t\times n_t}$ minimizing the approximate MSE $\hat{\mathcal{J}}_{\mathrm{vp},q}$ for cell-q with respect to the current channel coherence block is given by the n_t eigenvectors of $\mathbf{I}_{N_t} + \hat{\mathbf{R}}_{-q}^{-1}\hat{\mathbf{H}}_{q,q}^{\mathrm{H}}\hat{\mathbf{H}}_{q,q}$ corresponding to the n_t largest eigenvalues.*

Proof See Appendix 2.

Differently from the BD-based RF solutions in (6.24), the effect of channel tracking errors can be directly combined with the approximate VP-related MSE (6.28). In particular, the design of RF beamformer for the current channel fading block is formulated as

$$\begin{aligned} \min \ & \sum_{b\in\mathcal{Q}} \hat{\mathcal{J}}_{\mathrm{vp},b} + \omega_{\mathrm{kf}}\mathcal{E}_{\mathrm{kf}} \\ \text{s.t. } & \mathbf{F}_{\mathrm{RF},b}^{\mathrm{H}}\mathbf{F}_{\mathrm{RF},b} = \mathbf{I}_{n_t},\, b\in\mathcal{Q}, \end{aligned} \tag{6.29}$$

where $\{\mathbf{F}_{\mathrm{RF},b}\}_{b\in\mathcal{Q}}$ are the variables, and the channel tracking errors $\mathcal{E}_{\mathrm{kf}}$ resulting from Kalman filtering is given in (6.20). As mentioned in Sect. 6.3.2, the instantaneous effective CSI is not yet available in the design of the RF beamformer for the current transmission block, which is thus replaced by its MMSE prediction. This is indicated by $\{\hat{\mathcal{J}}_{\mathrm{vp},b}\}$ in place of $\{\mathcal{J}_{\mathrm{vp},b}\}$. Since the objective function with respect to each cell relies on cell-wise local CSI, it can be optimized in parallel. Unfortunately, when the effect of channel tracking errors $\mathcal{E}_{\mathrm{kf}}$ is incorporated into the RF design, it is not clear if a closed-form solution exists, and hence we resort to the numerical optimization framework developed in Sect. 6.3. In order to particularize the framework in Algorithm 6.3 to the distributed case (6.29), it can be shown that the Euclidean gradient with respect to $\mathbf{F}_{\mathrm{RF},q}$ consists in

$$\begin{aligned} \nabla_{\mathbf{F}_{\mathrm{RF},q}} \sum_{b\in\mathcal{Q}} \hat{\mathcal{J}}_{\mathrm{vp},b} = & -\frac{2\tau^2}{\pi e}\det\big(\hat{\boldsymbol{\Xi}}_q\big)^{-\frac{1}{K_q}} \\ & \times \left[\mathbf{I}_{N_t} - \big(\textstyle\sum_{b\in\mathcal{Q}\backslash q}\hat{\mathbf{H}}_{q,b}^{\mathrm{H}}\hat{\mathbf{H}}_{q,b}\big)\mathbf{F}_{\mathrm{RF},q}\big(\mathbf{F}_{\mathrm{RF},q}^{\mathrm{H}}\hat{\mathbf{R}}_{-q}\mathbf{F}_{\mathrm{RF},q}\big)^{-1}\mathbf{F}_{\mathrm{RF},q}^{\mathrm{H}}\right] \\ & \times \hat{\mathbf{H}}_{q,q}^{\mathrm{H}}\hat{\boldsymbol{\Xi}}_q^{-1}\hat{\mathbf{H}}_{q,q}\mathbf{F}_{\mathrm{RF},q}\big(\mathbf{F}_{\mathrm{RF},q}^{\mathrm{H}}\hat{\mathbf{R}}_{-q}\mathbf{F}_{\mathrm{RF},q}\big)^{-1}, \end{aligned}$$

and $\nabla_{\mathbf{F}_{\mathrm{RF},q}}\mathcal{E}_{\mathrm{kf}}$ is given by (6.23).

6.5 Illustrative Results and Discussions

In this section, we numerically evaluate the error performance in terms of bit error rate (BER) of the proposed two-timescale CSI-based hybrid precoding solutions with centralized MMSE-VP and distributed MMSE-VP, respectively. We are interested to know if it is advantageous to adapt the RF precoder to the imperfect high-dimensional CSI prediction/estimation instead of the channel spatial correlation. Moreover, since it has been observed in [23] that VP-based precoding schemes are sensitive to channel estimation errors, we ask if such an issue can be effectively alleviated by the two-stage hybrid precoding structure, where the dimension of the channel seen by the baseband is decreased in the presence of the RF stage. For these purposes, three baseline schemes are presented for comparison: (1) the statistical BD-based multi-cell hybrid precoder (denoted by Stat. BD) [4, 20], (2) the statistical IA-based multi-cell hybrid precoder (denoted by Stat. IA) [5], and (3) the imperfect CSI-based multi-cell fully digital precoder (denoted by Digital). In particular, the channel correlation-constrained RF precoder of Stat. BD aims at nulling the inter-cell interference while that of Stat. IA attempts to balance self-transmission with interference mitigation. In the case of distributed MMSE-VP, the fully digital precoder is obtained by adapting the robust baseband design to the high-dimensional channel instead of the RF-precoded effective channel. Equipped with the insight from Chap. 5 that nonlinear precoding at baseband provides a substantial performance gain over its linear counterpart, here, we assume that nonlinear MMSE-VP in place of the original linear precoding is employed in the hybrid baselines.

6.5.1 Simulation Setup

We consider a system configuration of $Q = 3$ adjacent hexagonal cells, where each BS, located at the center, is equipped with $N_\mathrm{t} = 64$ transmit antennas. The inter-BS distance is $2R_\mathrm{c} = 180\,\mathrm{m}$. There is a group of 6 cell-edge users associated with each cell, which are uniformly distributed in a circular area with a radius of $R_\mathrm{s} = 20\,\mathrm{m}$. The users clustered in the same area are assumed to experience the same transmit spatial correlation for the ease of comparison with other baselines. The mean angles of departure with respect to the local BS and neighboring BSs are $0°$ and $\pm 10°$, respectively, and the angle spread is $\Delta = 8.5°$. The path loss $\alpha_\mathrm{pl} = (1 + d_{b,q,i}/d_\mathrm{ref})^{-\delta_\mathrm{pl}}$ is computed from $d_{b,q,i}$, i.e., the distance between user-$u_{q,i}$ and the BS-b, a reference distance $d_\mathrm{ref} = 40\,\mathrm{m}$, and a path loss exponent of $\delta_\mathrm{pl} = 2.5$. The angular information and path loss are then used to generate the transmit correlation matrices according to (6.1). The variance of the observation noise is $\sigma_e^2 = 10^{-2}$. 16-QAM modulation is employed for transmission, and the modulo constant τ is set as four times the minimum distance between constellation symbols. We choose the parameters of Algorithm 6.3 as follows: $\rho = 10^{-4}$, $\alpha_{\text{lb}} = 10^{-20}$, $\alpha_{\text{ub}} = 10^{20}$, $\alpha = 10^{-3}$, $\eta = 0.85$, and $\delta = 0.1$ [21]. The channel coherence time is $T = 40$ channel uses. Following Jakes' model, we set $\xi_{q,i} = 0.9999, \forall q, i$ for simplicity, which corresponds to a carrier frequency of 2 GHz, a symbol duration of 0.1 ms, and mobile speeds less than 15 km/h. The signal-to-noise ratio (SNR) per user is defined as $QP_{\text{t}}/K\sigma_n^2$. The numerical results are averaged over 1000 Monte Carlo trials with 40 Kalman filtering iterations per trial.

6.5.2 *Performance Evaluation*

As shown in Fig. 6.3, when the centralized MMSE-VP is employed at baseband, the proposed hybrid precoding scheme (denoted by MSE (Predicted CSI)) remarkably outperforms the state-of-the-art baselines Stat. BD and Stat. IA whose RF components are adaptive to the channel transmit correlation. From the simulation setup, the power azimuth spectrum of the clustered users with respect to each cell is seen to overlap with each other, which translates to the overlapping of the corresponding transmit correlation matrices. Hence, it calls into question the feasibility to achieve good separation in the transmit correlation-constrained RF beam domain. In contrast, although not exactly the same as the true CSI, the high-dimensional predicted CSI produced by channel tracking provide more flexibility for generating the RF beamforming directions. Furthermore, the performance degradation caused by the inaccuracy of the channel predictions can be compensated by the baseband precoding, which is based on the instantaneous observations of the RF-beamformed channel. In that nonlinear VP-based techniques are sensitive to channel estimation errors [23], the estimation of a low-dimensional effective channel helps alleviate the adverse effect of the channel estimation errors. Such an intuition is verified in Fig. 6.3 by the performance advantage of the proposed hybrid design MSE (Predicted CSI)over the fully digital solution Digital, which is adaptive to the imperfect high-dimensional CSI.

As illustrated in Fig. 6.4, the previous observations also hold valid in the case of baseband distributed MMSE-VP. Interestingly, although the proposed BD scheme (denoted by BD (Predicted CSI)) only approximately nulls the inter-cell interference based on the predicted CSI, the residual interference is nonetheless sufficiently small to allow for cell-wise baseband precoding without suffering an interference-limited behavior in the operating SNR regime of interest. In this case, it delivers an almost identical BER as MSE (Predicted CSI). In Fig. 6.5, we further examine the impact of channel estimation errors, as represented by the error variance σ_e^2. For the ideal case where the perfect CSI ($\sigma_e^2 = 0$) is obtainable, it is seen that the performance of the proposed hybrid schemes, i.e., MSE and BD is indistinguishable from that of the fully digital solution. In other words, the addition of the RF beamforming

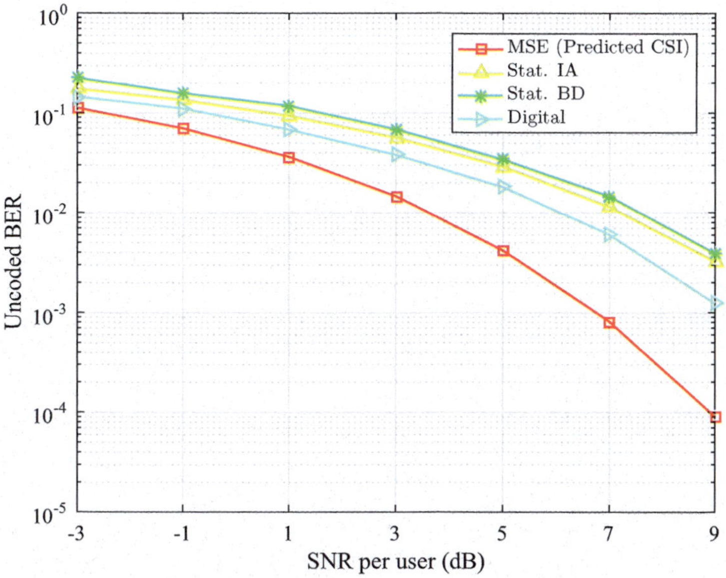

Fig. 6.3 BER curves of fully digital and hybrid precoding with centralized MMSE-VP, where three BSs, each equipped with 64 transmit antennas and 6 RF chains, serve a cluster of 6 cell-edge users within each cell

stage only causes negligible performance loss while enjoying significantly reduced hardware complexity. In the presence of minor channel estimation errors ($\sigma_e^2 = 10^{-2.5}$), the RF beamformer also helps reduce the exposure to the channel estimation errors at baseband, which thus suffers less severe performance degradation than the fully digital counterpart Digital. On the other hand, while all the presented schemes benefit from increased accuracy of channel estimation, as indicated by $\sigma_e^2 = 10^{-2}$ decreased to $\sigma_e^2 = 10^{-2.5}$, the single-stage solution Digital exhibits a greater extent of performance improvement than the two-stage hybrid solutions.

In Fig. 6.6, we investigate the CSI effect on the performance of the proposed hybrid precoding schemes, whose BERs are plotted for imperfect CSI, predicted CSI, and estimated CSI, respectively. Not surprisingly, the imperfect CSI-based solutions suffer the worse performance deterioration due to increased exposure to the channel estimation errors. Since Kalman filtering attempts to correct the channel predictions based on the partial observations of the channel, the MSE-based hybrid precoding design indeed benefits from a more accurate channel estimation. As the Kalman iteration progresses, the quality of the predicted CSI becomes gradually favorable, which in turn enhances the average error performance. This

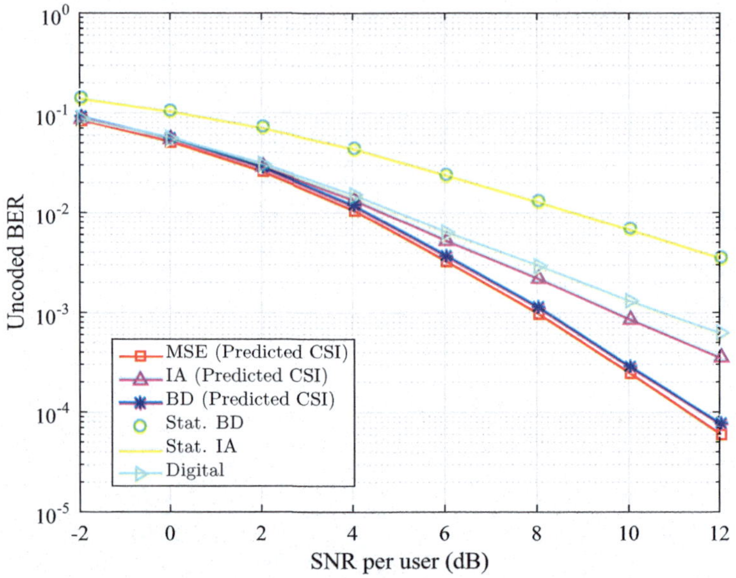

Fig. 6.4 BER curves of fully digital and hybrid precoding with distributed MMSE-VP, where three BSs, each equipped with 64 transmit antennas and 6 RF chains, serve a cluster of 6 cell-edge users within each cell

is confirmed by the small performance gap between MSE (Predicted CSI) and MSE (Estimated CSI).

Although the error performance of the proposed hybrid precoding solutions is only slightly degraded from using the CSI predictions, it relies on the initial availability of the imperfect high-dimensional CSI. However, such high-dimensional CSI is not always readily obtainable especially when the acquisition depends on the mechanism of channel training and feedback. We therefore examine the error performance of the hybrid precoding in the absence of initial CSI in Fig. 6.7. In this case, it is not longer sufficient to use the predicted CSI in the RF beamforming computation, as evidenced by the noticeable performance gap between the estimated CSI-based solutions, i.e., BD (Estimated CSI) and MSE (Estimated CSI)), and the predicted CSI-based counterparts, i.e., BD (Predicted CSI) and MSE (Predicted CSI). Interestingly, by comparing Figs. 6.6 and 6.7, we see that the availability of the initial channel estimation has only a minor impact on the MSE-based formulation but is critical to the performance of the approximate BD-based formulation.

6.6 Summary

In this chapter, we studied two-timescale CSI-based hybrid precoding with MMSE-VP for clustered cell-edge users in a multi-cell massive MIMO environment. Built

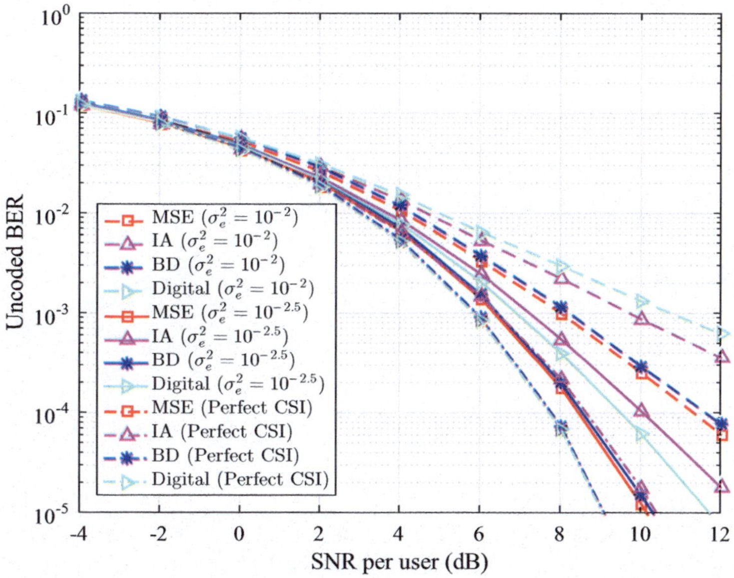

Fig. 6.5 BER curves of imperfect CSI-based fully digital solution and channel tracking-based hybrid precoding solutions with distributed MMSE-VP. The imperfect high-dimensional CSI is initialized with error variance $\sigma_e^2 = 10^{-2}$, $\sigma_e^2 = 10^{-2.5}$, and $\sigma_e^2 = 0$, respectively. Three BSs, each equipped with 64 transmit antennas and 6 RF chains, serve a cluster of 6 cell-edge users within each cell

upon the noisy observations of the RF-beamformed channel and the perfect knowledge of such statistical CSI as spatial and temporal correlation, Kalman filtering was exploited to estimate and track the instantaneous realization of the high-dimensional MIMO channel. Such channel estimates served as the basis to generate RF beamforming solutions in respect of centralized and distributed MMSE-VP at baseband. In particular, the nonlinear baseband solutions which were robust to the imperfect effective CSI were derived via low-complexity alternating optimization. Through abstracting the effect of the nonlinear MMSE-VP on the entire channel fading block, we arrived at a separate RF formulation, which was posed as solutions to balance the channel tracking accuracy with the system error performance. The combination of hybrid precoding with channel tracking was numerically justified by the noticeable error performance gain over other two-timescale CSI-based baselines, which constrained the RF components to the subspace of the transmit spatial correlation. Additionally, the proposed hybrid design effectively reduced the exposure to the channel estimation errors at baseband, and therefore enjoyed a superior BER to the high-dimensional CSI-based fully digital counterpart.

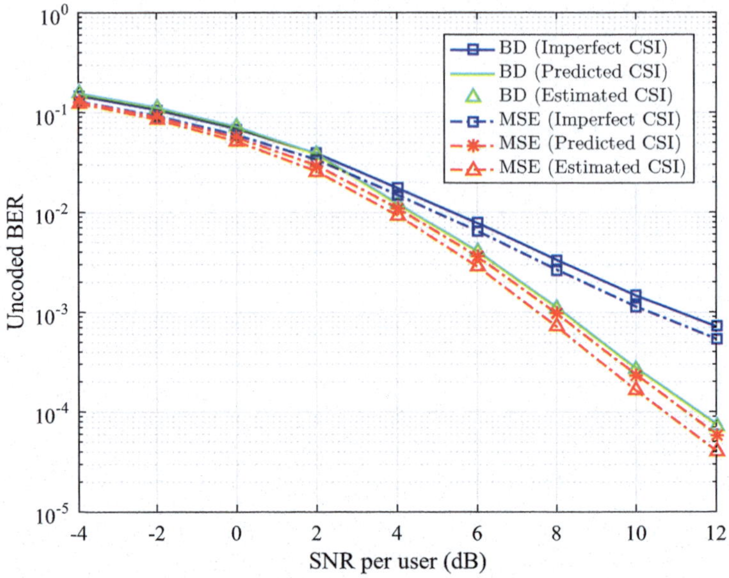

Fig. 6.6 BER curves of hybrid precoding with distributed MMSE-VP based on channel tracking and imperfect CSI. The imperfect high-dimensional CSI is initially available. Three BSs, each equipped with 64 transmit antennas and 6 RF chains, serve a cluster of 6 cell-edge users within each cell

Appendix 1: Proof of Theorem 6.1

By substituting in the expression of $\mathbf{K}_{b,q,i}$, and using the assumption that $\boldsymbol{\Sigma}_{b,b,1} = \cdots = \boldsymbol{\Sigma}_{b,b,K_b}$, it can be seen that minimization of the channel tracking errors is equivalent to

$$\max_{\mathbf{F}_{\mathrm{RF},b}^{[n]}} \operatorname{tr}\left[\left(\mathbf{F}_{\mathrm{RF},b}^{[n]\mathrm{H}}\tilde{\mathbf{P}}_{b,b,1}^{n|n-1}\mathbf{F}_{\mathrm{RF},b}^{[n]}\right)^{-1}\mathbf{F}_{\mathrm{RF},b}^{[n]\mathrm{H}}\left(\mathbf{P}_{b,b,1}^{n|n-1}\right)^{2}\mathbf{F}_{\mathrm{RF},b}^{[n]}\right]$$

$$\text{s.t. } \mathbf{F}_{\mathrm{RF},b}^{[n]\mathrm{H}}\mathbf{F}_{\mathrm{RF},b}^{[n]} = \mathbf{I}_{n_{\mathrm{t}}},\ b \in \mathcal{Q},$$

where $\tilde{\mathbf{P}}_{b,b,i}^{n|n-1} \triangleq \mathbf{P}_{b,b,i}^{n|n-1} + \sigma_e^2\mathbf{I}_{N_{\mathrm{t}}}$ is PD and thus invertible. Following [24], it can be concluded that an optimal solution to such a quotient trace formulation is given by the n_{t} eigenvectors of $(\tilde{\mathbf{P}}_{b,b,1}^{n|n-1})^{-1}(\mathbf{P}_{b,b,1}^{n|n-1})^2$ corresponding to the n_{t} largest eigenvalues.

Appendix 2: Proof of Theorem 6.2

We recall that $\hat{\mathbf{R}}_{-q} = \sum_{b\in\mathcal{Q}\setminus q}\hat{\mathbf{H}}_{q,b}^{\mathrm{H}}\hat{\mathbf{H}}_{q,b} + \xi_q\mathbf{I}_{N_{\mathrm{t}}}$ with $\{\hat{\mathbf{H}}_{q,b}\}_{b\in\mathcal{Q}}$ as the prediction for the current channel coherence block. Clearly, minimizing $\hat{\mathcal{J}}_{\mathrm{vp},q} =$

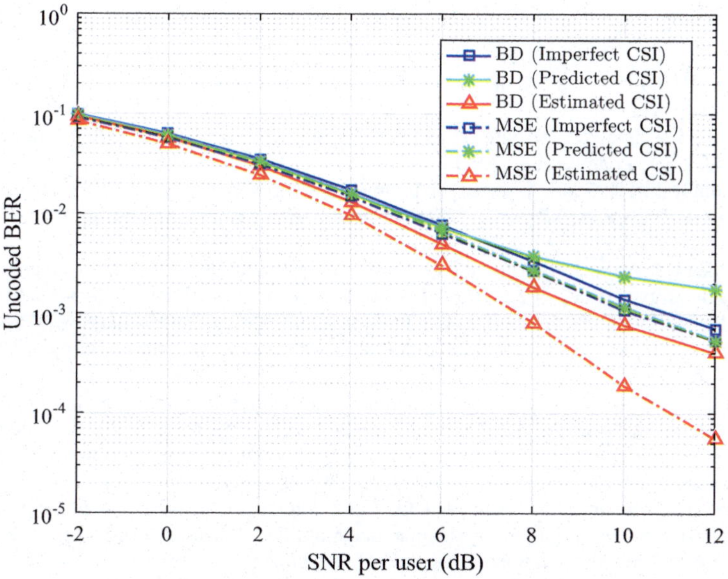

Fig. 6.7 BER curves of hybrid precoding with distributed MMSE-VP based on predicted CSI, estimated CSI, and imperfect CSI. The channel tracking is initiated from the zero state. Three BSs, each equipped with 64 transmit antennas and 6 RF chains, serve a cluster of 6 cell-edge users within each cell

$\frac{K_q\tau^2}{\pi e}\det(\hat{\boldsymbol{\Xi}}_q)^{1/K_q}$ is equivalent to minimizing $\det(\hat{\boldsymbol{\Xi}}_q)$. Using the identity $\det(\mathbf{I}+\mathbf{AB})=\det(\mathbf{I}+\mathbf{BA})$ and the assumption that $\mathbf{F}_{\mathrm{RF},q}^{\mathrm{H}}\mathbf{F}_{\mathrm{RF},q}=\mathbf{I}_{n_\mathrm{t}}$, the following chain of equalities can be established, i.e.,

$$\begin{aligned}\det(\hat{\boldsymbol{\Xi}}_q) &= \det\left[\mathbf{I}_{K_q}+\hat{\mathbf{H}}_{q,q}\mathbf{F}_{\mathrm{RF},q}\left(\mathbf{F}_{\mathrm{RF},q}^{\mathrm{H}}\hat{\mathbf{R}}_{-q}\mathbf{F}_{\mathrm{RF},q}\right)^{-1}\mathbf{F}_{\mathrm{RF},q}^{\mathrm{H}}\hat{\mathbf{H}}_{q,q}^{\mathrm{H}}\right]^{-1}\\ &= \det\left[\mathbf{I}_{n_\mathrm{t}}+\left(\mathbf{F}_{\mathrm{RF},q}^{\mathrm{H}}\hat{\mathbf{R}}_{-q}\mathbf{F}_{\mathrm{RF},q}\right)^{-1}\mathbf{F}_{\mathrm{RF},q}^{\mathrm{H}}\hat{\mathbf{H}}_{q,q}^{\mathrm{H}}\hat{\mathbf{H}}_{q,q}\mathbf{F}_{\mathrm{RF},q}\right]^{-1}\\ &= \left[\frac{\det[\mathbf{F}_{\mathrm{RF},q}^{\mathrm{H}}(\hat{\mathbf{R}}_{-q}+\hat{\mathbf{H}}_{q,q}^{\mathrm{H}}\hat{\mathbf{H}}_{q,q})\mathbf{F}_{\mathrm{RF},q}]}{\det\left(\mathbf{F}_{\mathrm{RF},q}^{\mathrm{H}}\hat{\mathbf{R}}_{-q}\mathbf{F}_{\mathrm{RF},q}\right)}\right]^{-1}.\end{aligned}$$

This is the generalized Rayleigh quotient formulation [25], to which a specific solution is given by the n_t eigenvectors of $\hat{\mathbf{R}}_{-q}^{-1}(\hat{\mathbf{R}}_{-q}+\hat{\mathbf{H}}_{q,q}^{\mathrm{H}}\hat{\mathbf{H}}_{q,q})$, i.e., $\mathbf{I}_{N_\mathrm{t}}+\hat{\mathbf{R}}_{-q}^{-1}\hat{\mathbf{H}}_{q,q}^{\mathrm{H}}\hat{\mathbf{H}}_{q,q}$, corresponding to the n_t largest eigenvalues.

References

1. Huawei, "R1-050507: Soft frequency reuse scheme for UTRAN LTE," *3GPP TSG RAN WG1 Meeting #41*, May 2005.
2. A. Ghosh, J. Zhang, J. G. Andrews, and R. Muhamed, *Fundamentals of LTE*. Upper Saddle River, NJ, USA: Prentice Hall, 2010.
3. J. G. Andrews, R. K. Ganti, M. Haenggi, N. Jindal, and S. Weber, "A primer on spatial modeling and analysis in wireless networks," *IEEE Commun. Mag.*, vol. 48, no. 11, pp. 156–163, Nov. 2010.
4. A. Adhikary, J. Nam, J.-Y. Ahn, and G. Caire, "Joint spatial division and multiplexing - The large-scale array regime," *IEEE Trans. Inf. Theory*, vol. 59, no. 10, pp. 6441–6463, Oct. 2013.
5. J. T. Chen and V. K. N. Lau, "Two-tier precoding for FDD multi-cell massive MIMO time-varying interference networks," *IEEE J. Sel. Areas Commun.*, vol. 32, no. 6, pp. 1230–1238, Jun. 2014.
6. D. Kim, G. Lee, and Y. Sung, "Two-stage beamformer design for massive MIMO downlink by trace quotient formulation," *IEEE Trans. Commun.*, vol. 63, no. 6, pp. 2200–2211, Jun. 2015.
7. Y. Xu, G. Yue, and S. Mao, "User grouping for massive MIMO in FDD systems: New design methods and analysis," *IEEE Access*, vol. 2, pp. 947–959, 2014.
8. J. Nam, A. Adhikary, J.-Y. Ahn, and G. Caire, "Joint spatial division and multiplexing: Opportunistic beamforming, user grouping and simplified downlink scheduling," *IEEE J. Sel. Topics Signal Process.*, vol. 8, no. 5, pp. 876–890, Oct. 2014.
9. J. Nam, Y. J. Ko, and J. Ha, "User grouping of two-stage MU-MIMO precoding for clustered user geometry," *IEEE Commun. Lett.*, vol. 19, no. 8, pp. 1458–1461, Aug. 2015.
10. D.-S. Shiu, G. J. Foschini, M. J. Gans, and J. M. Kahn, "Fading correlation and its effect on the capacity of multielement antenna systems," *IEEE Trans. Commun.*, vol. 48, no. 3, pp. 502–513, Mar. 2000.
11. S. Noh, M. D. Zoltowski, Y. Sung, and D. J. Love, "Pilot beam pattern design for channel estimation in massive MIMO systems," *IEEE J. Sel. Topics Signal Process.*, vol. 8, no. 5, pp. 787–801, Oct. 2014.
12. B. M. Hochwald, C. B. Peel, and A. L. Swindlehurst, "A vector-perturbation technique for near-capacity multiantenna multiuser communication - Part II: Perturbation," *IEEE Trans. Commun.*, vol. 53, no. 3, pp. 537–544, Mar. 2005.
13. L. Grippo and M. Sciandrone, "On the convergence of the block nonlinear Gauss-Seidel method under convex constraints," *Operations Research Lett.*, vol. 26, no. 3, pp. 127–136, Apr. 2000.
14. E. Agrell, T. Eriksson, A. Vardy, and K. Zeger, "Closest point search in lattices," *IEEE Trans. Inf. Theory*, vol. 48, no. 8, pp. 2201–2214, Aug. 2002.
15. D. A. Schmidt, M. Joham, and W. Utschick, "Minimum mean square error vector precoding," *European Trans. Telecommun.*, vol. 19, no. 3, pp. 219–231, Apr. 2008.
16. A. Kammoun, A. Muller, E. Bjornson, and M. Debbah, "Linear precoding based on polynomial expansion: Large-scale multi-cell MIMO systems," *IEEE J. Sel. Topics Signal Process.*, vol. 8, no. 5, pp. 861–875, Oct. 2014.
17. Y. Yang and M. Pesavento, "A unified successive pseudoconvex approximation framework," *IEEE Trans. Signal Process.*, vol. 65, no. 13, pp. 3313–3328, Jul. 2017.
18. J. H. Conway and N. J. A. Sloane, *Sphere Packings, Lattices and Groups*, ser. Grundlehren der mathematischen Wissenschaften. New York, NY, USA: Springer-Verlag, 2013.
19. D. J. Ryan, I. B. Collings, I. V. L. Clarkson, and R. W. Heath, "Performance of vector perturbation multiuser MIMO systems with limited feedback," *IEEE Trans. Commun.*, vol. 57, no. 9, pp. 2633–2644, Sep. 2009.
20. A. Adhikary and G. Caire, "JSDM and multi-cell networks: Handling inter-cell interference through long-term antenna statistics," in *Proc. IEEE 48th Asilomar Conf. on Signals, Systems and Comput.*, Pacific Grove, CA, USA, Nov. 2014.

21. Z. Wen and W. Yin, "A feasible method for optimization with orthogonality constraints," *Mathematical Programming*, vol. 142, no. 1, pp. 397–434, Dec. 2013.
22. Q. H. Spencer, A. L. Swindlehurst, and M. Haardt, "Zero-forcing methods for downlink spatial multiplexing in multiuser MIMO channels," *IEEE Trans. Signal Process.*, vol. 52, no. 2, pp. 461–471, Feb. 2004.
23. M. Mazrouei-Sebdani, W. A. Krzymień, and J. Melzer, "Massive MIMO with nonlinear precoding: Large-system analysis," *IEEE Trans. Veh. Technol.*, vol. 65, no. 4, pp. 2815–2820, Apr. 2016.
24. S. C. Yan and X. O. Tang, "Trace quotient problems revisited," in *Proc. 9th European Conf. Comput. Vision*, Grazm, Austria, May 2006.
25. R. E. Prieto, "A general solution to the maximization of the multidimensional generalized Rayleigh quotient used in linear discriminant analysis for signal classification," in *Proc. IEEE Int. Conf. Acoustics, Speech and Signal Process.*, Hong Kong, China, Apr. 2003.

Chapter 7
Conclusions

7.1 Integration of Hybrid Precoding with C-RAN

In the previous chapter, the ideas of centralized and distributed baseband precoding in the hybrid RF-baseband architecture were explored for the purpose of suppressing the inter-cell interference (ICI) and improving the system error performance at the cell edge. In the distributed implementation, it is seen that by partitioning the high-dimensional multiple-input multiple-output (MIMO) channel into multiple near-orthogonal, low-dimensional subspaces as dictated by block diagonalization in the RF domain, cell-wise spatial multiplexing can be carried out at baseband without causing severe ICI. Although this obviates the need of channel state information (CSI) and user data exchange between base stations (BSs), the lack of cell coordination unfortunately prevents balancing intra-cell spatial multiplexing and ICI suppression in a resource efficient manner. On the contrary, by allowing inter-BS collaboration as in the centralized approach to baseband spatial multiplexing, the error performance enjoys a significant enhancement.

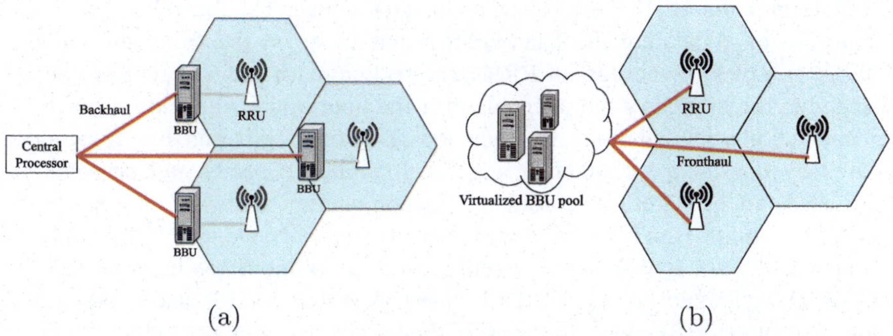

Fig. 7.1 Conventional inter-BS connection via backhauling & C-RAN. (**a**) Conventional RAN. (**b**) C-RAN

T. Le-Ngoc, R. Mai, *Hybrid Massive MIMO Precoding in Cloud-RAN*, Wireless Networks, https://doi.org/10.1007/978-3-030-02158-0_7

In traditional radio access networks (RANs), each macro-cell is serviced by a full-fledged BS, which comprises a remote radio head (RRH) and a baseband unit (BBU), as shown in Fig. 7.1a. To enable coordination and collaboration between cells, message exchange between BSs takes place over backhaul connections, be it fiber optic or wireless. For example, the mechanism of backhauling has been defined via the Xn interface for macro-cells in 3GPP 5G New Radio (NR) Release 15 [1]. Since the macro-cell BSs are in general geographically separated apart, tight synchronization as required by coordinated beamforming/scheduling and joint processing poses a nontrivial challenge. Furthermore, the computational resources of the BBU constitute a limiting factor, among others, to the capacity of the cell. In fact, study has shown that due to the non-uniform nature of data traffic, the radio resources are not sufficiently exploited by such a setup. In an effort to address the drawbacks of traditional backhauling and further enhance the performance of radio resource management, the architecture of C-RAN has been proposed [2, 3].

As illustrated in Fig. 7.1b, a characteristic element of C-RAN is the centralization and virtualization of the BBUs of a cluster of BSs, which are connected to the separate RRHs via the fronthaul transport network. In its simplest form, the RRHs are only concerned with the transmission and reception of radio signals while sophisticated signal processing at the physical layer such as up-/down-conversion, analog-to-digital/digital-to-analog conversion (ADC/DAC), modulation/demodulation, Fast Fourier Transform (FFT)/Inverse FFT (IFFT) is taken care of by the BBUs. In doing this, the computational and storage resources of the virtualized BBU pool can be adaptively allocated based on the traffic demand from each of the clustered cells. The resulting statistical multiplexing gain in turn translates to a saving on the number of BBUs in deployment and an improvement in energy efficiency [4].

Depending on the functional split, the transport network between the RUs and the virtualized BBU pool can be realized in both digital and analog forms. In particular, the digital interface, i.e., Common Public Radio Interface (CPRI), has been widely adopted for the transfer of I/Q digital signals. For this purpose, the functionalities of mixing and ADC/DAC reside in the RRHs, and strict synchronization between the RRHs and the BBU pool needs to be guaranteed. On the other hand, for the purpose of alleviating the bandwidth requirement on the fronthaul and the hardware complexity and cost of RRHs, the technique of radio over fiber can be leveraged to directly carry radio signals over the fiber optic fronthaul. In practical implementations, the interplay between the functional split and the constraints imposed by non-ideal fronthauling, e.g., with limited capacity and propagation delay, needs to be carefully considered in conjunction with the chosen signal processing techniques in order to reap the performance advantage of C-RAN [4].

Conceptually, the hybrid RF-baseband precoding solutions can be implemented in a C-RAN architecture where the RRUs are tasked with the RF analog precoding and the virtualized BBU pool takes care of the baseband digital precoding. Enabled by the centralized computational resources in the cloud, the RF beamforming weights are first derived in the BBU pool, and then are used to update the beamformers of the RRUs. In the presence of fast fading MIMO channels, the processing delay and the propagation delay introduced by the non-ideal fronthaul transport network are likely to render the RF beamforming weights obsolete. Fortunately, the two-timescale CSI-based hybrid precoding schemes are well-suited to address this concern, since the statistical CSI, which the RF precoding is built upon, remains approximately constant over a relatively long period of time.

The next-generation cellular networks are evolving toward the ultra-dense paradigm where micro-/femto-/pico-cells might be added to extend cell coverage, expand network capacity, and off-load user traffic. This is facilitated with the coordination of the macro-cells, where signaling can take place over the mmWave backhauling to achieve low latency. In this use case, the joint design of two-timescale hybrid precoding and combining for point-to-point massive MIMO wireless backhauling subject to the statistical queueing constraint in Chap. 3 might prove useful. When a user experiences a poor link quality as a result of severe path loss or strong inter-user interference, a successful delivery of data packets at a target rate is not always possible. As a remedy, hybrid automatic repeat request (ARQ)-enabled packet retransmission can be used to improve the link reliability. To this end, the hybrid precoding designs exploiting the temporal diversity inherent in packet retransmissions can serve as a potential solution.

The most recent 5G NR Release 15 has prescribed a more uniform user experience than its predecessor [1], which implies a substantial rate improvement at the cell edge. To alleviate or even eliminate the adverse effect of ICI, multi-cell transmission can be frequency multiplexed. For example, the control and data channels across interfering cells can be assigned an orthogonal set of subcarriers in orthogonal frequency division multiplexing (OFDM) systems. Over the allocated subcarriers, it is desirable to spatially multiplex as many users as possible in order to achieve high area spectral efficiency. In view of the severe performance degradation suffered by the linear precoding schemes such as zero-forcing and regularized zero-forcing in fully loaded homogeneous multi-user MIMO systems, the use of nonlinear vector perturbation (VP) technique in a two-timescale hybrid minimum mean square error (MMSE) precoding design was demonstrated to be an encouraging option in Chap. 5. One advantage of C-RAN is the significantly reduced delay and rich computational resources aggregated at baseband across the clustered cells. Therefore, extension of the nonlinear hybrid precoding technique can be made to consider joint transmission in a multi-cell environment, as done in Chap. 6.

7.2 Future Work

The research subject of this monograph, i.e., hybrid RF-baseband precoding/combining, represents a potential avenue to overcome the technical roadblocks that lie in the migration from conventional MIMO to massive MIMO. To conclude this monograph, we wish to point out research directions that are worth further exploration towards the practical implementation and standardization of massive MIMO:

1. *Hybrid precoding with imperfect CSI:* For the two-timescale hybrid RF-baseband precoding/combining designs, the high-dimensional channel correlation matrix has been assumed to be perfectly known at the BS. For the one-ring channel model used in Chaps. 5 and 6, the estimation of such second-order channel statistics boils down to the estimation of the mean angle of departure and angle spread. This in turn relies on uplink channel training. In general, the channel spatial correlation is rank deficient in a limited scattering environment such as millimeter-wave communications. It is therefore of practical interest to study how to exploit such a low-rank property in the design of efficient channel training techniques [5, 6]. Furthermore, the current standardization efforts by 3GPP 5G NR mainly focus on digital CSI feedback. So far, we have assumed that the analog feedback of the effective channel coefficients is available. One may be interested to know how to adapt such a scheme to quantized effective CSI, where it becomes necessary to take into account the presence of RF beamforming in the baseband codebook construction.
2. *Hybrid precoding with finite-resolution RF elements:* The techniques proposed in this research have built upon the assumption that the magnitude and phase of an RF element can be arbitrarily adjusted. In general, the RF beamforming weights are computed at baseband and then uploaded to the RF domain. In light of a large number of high-resolution phase-shifters employed for a large-scale antenna array, such uploading demands a high-bandwidth data link. This motivates one to ask how to optimize the hybrid precoder when the RF beamformer is limited to a finite resolution. Preliminary efforts were made in Chap. 4, where the performance of RF precoding/combining with uniformly quantized elements was numerically evaluated. Besides, in order to guarantee the linearity and efficiency of power amplifiers, the operating point of power amplifiers can be fixed in practice, which translates to the modeling of RF elements as constant-modulus. Currently, the proposal of massive MIMO in 3GPP LTE cellular systems considers the feedback of RF beam indices to facilitate CSI estimation and reporting. Therefore, it has practical implications to investigate the optimal codebook construction for RF phase-shifting [7].
3. *Hybrid precoding/combining for in-band full-duplex massive MIMO systems:* By allowing simultaneous transmission and reception, the full-duplex operation mode promises to double the spectral efficiency of the traditional half-duplex systems [8]. However, because of the stark disparity between the transmit and receive power, the strong interference from the transmitter, i.e., self-interference, causes the analog-to-digital converters at the receiver to quickly become saturated, and calls into question the feasibility of full-duplex systems. Fortunately, the decreased path loss as a result of BS densification and the high beamforming gain enabled by massive MIMO provide a possibility for low-power transmission, which hence significantly mitigates the overwhelming self-interference [9]. It remains to be seen how to exploit the largely available spatial degrees of freedom for self-interference cancellation together with spatial multiplexing in the presence of the hybrid RF-baseband precoding/combining structure [10].

References

1. E. Dahlman, S. Parkvall, and J. Sköld, *5G NR: The Next Generation Wireless Access Technology*. Cambridge, MA, USA: Academic Press, 2018.
2. Y. Lin, L. Shao, Z. Zhu, Q. Wang, and R. K. Sabhikhi, "Wireless network cloud: Architecture and system requirements," *IBM J. Res. Develop.*, vol. 54, no. 1, pp. 4:1–4:12, Jan./Feb. 2010.
3. C. China Mobile Research Institute, Beijing, "C-RAN: The road towards green RAN," *Tech. Rep.*, Oct. 2011.
4. A. Checko, H. L. Christiansen, Y. Yan, L. Scolari, G. Kardaras, M. S. Berger, and L. Dittmann, "Cloud RAN for mobile networks–A technology overview," *IEEE Commun. Surveys Tuts.*, vol. 17, no. 1, pp. 405–426, First Quarter 2015.
5. Y. Wang, Z. Tian, S. L. Feng, and P. Zhang, "Efficient channel statistics estimation for millimeter-wave MIMO systems," in *Proc. IEEE Int. Conf. Acoustics, Speech and Signal Process.*, Shanghai, China, Mar. 2016.
6. J. Fang, X. J. Li, H. B. Li, and F. F. Gao, "Low-rank covariance-assisted downlink training and channel estimation for FDD massive MIMO systems," *IEEE Trans. Wireless Commun.*, vol. 16, no. 3, pp. 1935–1947, Mar. 2017.
7. K. Venugopal, A. Alkhateeb, N. G. Prelcic, and R. W. Heath, "Channel estimation for hybrid architecture-based wideband millimeter wave systems," *IEEE J. Sel. Areas Commun.*, vol. 35, no. 9, pp. 1996–2009, Sep. 2017.
8. D. Bharadia, E. McMilin, and S. Katti, "Full duplex radios," in *Proc. ACM Special Interest Group on Data Commun.*, Hong Kong, China, Aug. 2013.
9. A. Sabharwal, P. Schniter, D. N. Guo, D. W. Bliss, S. Rangarajan, and R. Wichman, "In-band full-duplex wireless: Challenges and opportunities," *IEEE J. Sel. Areas Commun.*, vol. 32, no. 9, pp. 1637–1652, Sep. 2014.
10. R. Mai, D. H. N. Duy, and T. Le-Ngoc, "Joint MSE-based hybrid precoder and equalizer design for full-duplex massive MIMO systems," in *Proc. IEEE Int. Conf. Commun.*, Kuala Lumpur, Malaysia, May 2016.

Printed by Printforce, the Netherlands